Use R!

Series Editors:
Robert Gentleman   Kurt Hornik   Giovanni Parm

More information about this series at http://www.springer.com/series/6991

Use R!

Cedric Gondro

Primer to Analysis of Genomic Data Using R

 Springer

Cedric Gondro
Ctr. Genetic Analysis and Applications
University of New England
Armidale, NSW, Australia

ISSN 2197-5736   ISSN 2197-5744   (electronic)
Use R!
ISBN 978-3-319-14474-0   ISBN 978-3-319-14475-7   (eBook)
DOI 10.1007/978-3-319-14475-7

Library of Congress Control Number: 2015934220

Springer Cham Heidelberg New York Dordrecht London

Printed on acid-free paper

Springer International Publishing AG Switzerland is part of Springer Science+Business Media (www.
springer.com)

To Placido, Walburga, Simone and Sean

Preface

Overview

Just about any text written on the analysis of genomic data will begin by mentioning the rapid pace of changes in the field. How the technology is frantically moving forward and how datasets are getting bigger and bigger. A huge experiment one year is just a tiny *proof of concept* the following year. Databases are growing exponentially. The literature on even quite specific subjects is overwhelming and we have to decide if we are going to keep up to date or actually get some of the work done.

It feels that just a few years ago, a genome scan with 300–400 microsatellite markers was a pretty big deal (in truth, it really was just a few years ago)! Then along came the 10k SNP chip, then the 50k, the 500k, and then, of course, the one, two, and three million SNP arrays. Full individual sequence data is rapidly becoming the platform of choice. At *10x* coverage, that's around 30 billion nucleotide reads per patient/animal/sample in the unprocessed *fastq* files.

Of course, we cannot manually operate at this kind of scale anymore. Data analysis became heavily dependent on computers and efficient algorithms to sift through the sea of data and make sense out of it all.

A plethora of computational tools have been written to cope with this high volume of data. Most of these have been developed to tackle specific problems, and even if they excel in their specific task, they may not be ideal for automated processes—the output of one tool is not in an adequate format for another tool further down the analysis pipeline. This leaves us with the task of finding out which tools are available for each step in an analysis, choosing the ones that meet our needs, figuring out how each one works, and sewing them together. Alternatively, some software (usually commercial, a.k.a. costs money) will seamlessly handle a full analysis from beginning to end, but the user is restricted to the choice of algorithms coded into the program, there's less flexibility in what can be done, and there's always a lingering feeling of *black box* about it. In recent years, R [90], a statistical programming language and environment, has become popular for the analysis of genomic data and even further has become the *de facto* tool for the analysis of gene expression data. R provides an integrated development environment for analysis and at the same time flexibility and full control of the analytic workflow.

In this book, we will focus on using R for the analysis of genomic data and how to set up routines to automate the analytical steps. We will not cover all that R can do (that in itself would be a rather large book and there are some very good ones already), but we will focus on some of the key points relevant to the analysis of genomic data: less emphasis on the theory and more emphasis on a practical hands-on, how to get the job done approach. The purpose of this book is to serve as a companion text for advanced undergraduate and graduate units in genomic analysis and bioinformatics and can be used as the practical component in *lab sessions*. The book should also be of use to researchers who want to use R for the analysis of genomic data.

Strictly speaking, no previous knowledge of R is necessary—the first chapter covers some of the basics, but readers will definitely benefit from some prior exposure to R. Familiarity with undergraduate level biostatistics and genetics is assumed.

I am mainly involved with livestock projects, so there is a very strong bias toward examples in this area. This is probably not such a bad thing since livestock lies between the genetics of humans, mice, and a few other model organisms for which the computational tools and sources of information available are quite mature, while on the other extreme we have species for which there are no commercial platforms, limited biological information in the databases, and none of the tools can be used *out of the box*. Livestock, e.g., cattle, sheep, and pigs, are an excellent resource to work on because we are forced to adapt methodologies to fit our needs, but at the same time we have sufficient resources to explore most of the facets at the cutting edge of genomic research.

What This Book *Is* About

- Understanding the basics of R needed to analyze genomic data
- R packages available for genomic analysis
- How to work with and manipulate large data files
- Building and using databases to store and retrieve data
- The importance of preprocessing and quality control with genomic data
- Principles of genome-wide association studies
- Principles of genomic prediction
- Principles of gene expression analysis
- Working with public databases to extract biological information
- Principles of population genetics
- Automation of analyses
- How to speed up R for high-throughput analysis
- How to use stand-alone software as part of an analysis pipeline in R

What This Book *Is Not* About

- A comprehensive overview of R syntax and programming
- Theoretical biostatistics
- Animal breeding
- And of course, as the title suggests, we will only scratch the surface of each topic

A Note on R

While this book is static (at least until the next edition), R is dynamic. Things can change rather quickly in R (but that's also true for the gene technologies). Add-on packages can change quite significantly from one version to the next, they might be replaced by another package, or they may simply not be maintained anymore. This constant state of flux can be aggravating at times, particularly when you have a long and complex analysis script that breaks down under a new version. And of course, these changes will do no favors to the book. Over time some things are bound to break down. Keep this in mind as you read/work your way through the book, and remember that the web is your friend to help figure out what went wrong because new versions were released.

This book was prepared using R version 3.1.1 and packages from Bioconductor 2.14. The focus is on the Windows version of R, but the examples should port to Linux or Mac without the need for any changes.

Data

All datasets used throughout the book can be downloaded from the publisher's website. There's one folder for each chapter, so it should be quite straightforward to match them together. Many R books provide accompanying datasets as R packages. I have decided to provide raw and untreated data in the exact same format as I routinely encounter. One of the key points of this book is on how to handle genomic data and this is best achieved by working sequentially through each step of the process and discussing the common data cleaning issues and pitfalls. The downside of this is that there are some rather lengthy data cleaning steps and readers will find it hard to skip around chapters without performing the previous steps.

Analysis of genomic data is computationally intensive in whatever way you look at it. And this of course is just a way of saying that you'll be spending more time in front of the monitor waiting for results than you would like to. Efficient programming can go a long way to reduce these *waiting periods*, but the computational resources available will also play a key role in the process—with a lot of processors, jobs can be parallelized; with a lot of memory, there will be no need to shuffle data between memory and disk. Unfortunately efficient programming ultimately is linked to computational resources. The best solution for a high-end machine might not be adequate for more modest resources; the job might simply run out of memory and crash. Here I have sought a balance that favors lower end resources at the expense of efficiency: any current personal desktop will run these examples in a reasonable time frame. From time to time throughout the chapters, we will also discuss some more efficient approaches in case the computational grunt is available.

Acknowledgments

Foremost, I would like to thank all the students who took the *Genomic Analysis* course with me; your feedback and comments on preliminary versions of this book have greatly improved its content and structure. Thanks also to Brian Kinghorn, John Gibson, Julius van der Werf, and Sam Clark, my colleagues, and friends at UNE for the many fruitful discussions we have had over the years—some sections of this book are based on notes developed by them. Thanks also to Laercio Porto-Neto and Hawlader Al Mamun for permission to use some text written by them. I also acknowledge funding support for this work from the Next-Generation BioGreen 21 Program (No. PJ009954), Rural Development Administration, Republic of Korea, and the Australian Research Council (DP130100542).

Armidale, NSW, Australia   Cedric Gondro

Contents

List of Figures

Chapter 1
R Basics

In this chapter we will cover the basic steps for getting started in R. We will discuss the pros and cons of R, how to install the software and additional packages, and some suggestions on how to set up the machine to use R efficiently. We will also see how to read, manipulate, summarize, plot, and save data—the cornerstones of any analysis.

1.1 Why R?

Before praising R's many virtues a short overview is in order. R is a software environment and programming language for statistical analysis. It is similar to the S language and a lot of code written in S can be used straight with R. Originally written by Robert Gentleman and Ross Ihaka from the University of Auckland, R is currently developed by the R Development Core Team [90]. At the time of writing the current version is 3.1.1.

In recent years R has become the *de facto* choice of many statisticians and is widely used to teach statistics courses at universities. Dozens of books have been published about R itself or on the use of statistical methodologies which are illustrated using the R environment. This book is part of a series published by Springer called *Use R!* which already has over 50 books published.

Now, going back to the title of this section, *why R?* For starters, it's free. Of course cost should never be the only determinant, but then again, R is free! The concept of free extends beyond just cost. R is free to use, is free to modify, is open source, and is platform free. What this essentially means is that it is easy to work

---

Electronic supplementary material The online version of this chapter (doi: 10.1007/ 978-3-319-14475-7_1) contains supplementary material, which is available to authorized users.

© Springer International Publishing Switzerland 2015
C. Gondro, *Primer to Analysis of Genomic Data Using R*, Use R!,
DOI 10.1007/978-3-319-14475-7_1

across platforms (e.g., Windows, Mac OS, and Linux), you can embed R into your own applications and you can change the code to suit your needs. R is released under the GNU public license.

Since R is a programming environment (and a programming language in its own right) users can write their own code to address particular needs without being restricted to predetermined types of analyses. Of course this comes at the cost of having to learn the syntax of the language—even though some point and click graphical interfaces have been developed, for example R Commander [36] which wraps R in a graphical environment for simple statistical analyses or AffyPipe (details in Chap. 7) for Affymetrix microarray analyses.

R consists of a base installation and can be extended through packages (somewhat the equivalent of programming libraries). There are thousands of packages available to tackle a wide range of problems. These packages are developed independently from the core program and can be downloaded from central repositories or, less frequently, directly from the developers' web sites. And this is probably the key feature of R. Chances are that someone has already written a package that will do what you need, saving hours or days of programming it yourself. Packages can be used together allowing the output of a function in one package to be used as the input for a function in another package—essentially you have at your fingertips an overwhelming set of building blocks. Just to illustrate, this book was written entirely in R.

And this brings us to why use R for analysis of genomic data. There are hundreds of packages specifically available for this task. There are packages for importing a wide range of data formats, preprocessing data, performing quality control tests, for the actual analytical steps, downstream integration with biological databases, and so on. A large number of new algorithms and methods are published and released as an R package at the same time, thus providing quick access to the most current technologies without the slow turnover time of commercial software. Of course, there's the risk that you will be the first to find out that the new hot algorithm is not really so cool!

But it's not all joy; the learning curve for R is pretty steep. The syntax, well, consider that packages are freely contributed by hundreds of developers across the world, there are no formal naming conventions and R is case sensitive! I'm sure you get the picture. For example, if you want to plot a heatmap it's anyone's guess if the syntax is *Heatmap, heatmap, HeatMap, HEATMAP* or, why not, *heat.Map*. And to add insult to injury many packages implement similar functions, such as the following real heatmap examples: *hclusterPlot, matrixPlot, heatmap, heatmap.2, heatmap_2, heatmap_plus, heatplot*, and so on. So it can become quite taxing to remember the name of the package, the name of the function in the package, and the proper casing. This of course does not sit well with those who, such as myself, by noon do not seem to remember quite well what they had for breakfast. Of course there are always the help files, but they are usually unhelpful, unless you know the exact name of the function you are looking for (don't believe me? Try to search for how to invert a matrix). And web searches have to be carefully worded or they will not be of great assistance either—have you ever thought of how non-informative the letter *R* is for searching the web? As a tip, when searching online try *r statistics*—that helps. Searches with *"how to ... using R"* also usually return meaningful results.

Of more practical importance is that genomic datasets have become extremely large and R was not designed for memory efficiency. There is however a strong drive in the R community to develop tools to handle these ever increasing in size datasets and new approaches/packages are continuously being developed. It is frequently intractable to load the entire dataset into memory and forget about using 32-bit Windows with its around 2 GB limit per process regardless of how much memory you have available. Some workarounds for this dimensionality problem are discussed in Chap. 3 but they will impact negatively on runtimes.

Another common comment is that R is slow, given that it is a scripted language and not a compiled one. This is not necessarily true since computationally intensive operations are usually written in C or Fortran and linked from within R. Of course we can be pedantic and say that *real R* is slow, but the practical result is that R is (or could be) just as fast as C or Fortran because that is what's running under the hood. Even though we are now in the *criticise R* section, I should highlight that since we can dynamically link to code in C or Fortran (and also other languages to various degrees), this opens the possibility of (1) using prior code or (2) developing code specifically tailored for solving a computationally intensive task and sending the results back into R to make use of its resources (e.g., plotting—R excels at it).

Time to get started with R...

1.2 Installing R

You can download the source code for R and compile it yourself or download binaries for quite a few platforms available from http://www.r-project.org/. Here we will focus on the Windows release, but most of the examples we will cover can be run on any platform without changes.

To install, download the executable (around 40 MB), double click to start it up, and then click on the usual *next, next, next ...* with a couple of options in between. The current version installs two versions of the R executable, a 32-bit and a 64-bit version if you are running a 64-bit version of Windows (only the 32-bit version is installed on 32-bit Windows). If you are using Mac OS, the R binary also installs 32- and 64-bit versions. Once installation is complete, to open R find the R folder in the *Start Menu* and click on R or R x64. This will open the R console (Fig. 1.1) and you are ready to go.

But before we start analyzing data, a couple of handy tips should be mentioned. If you right click on the shortcut to R and then on properties you will get the window shown in Fig. 1.2. You will see that in the path window a flag was added to increase memory, you'll really only need this if you are running the 32-bit version of R (the 64-bit version uses all the machine's memory by default). For example to increase

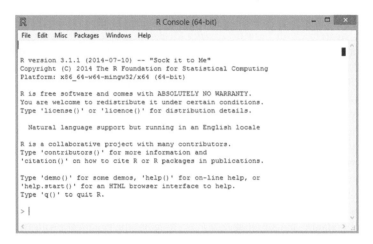

Fig. 1.1 Screenshot of R console

memory to 4G (maximum in R32 running Windows64) add to the path *--max-mem-size=4000M*. On shared resources, this flag can also be used to limit the maximum memory available to R.

You will also want to be able to access the internet from R (e.g., to install new packages). This may not be a problem, but many companies/universities use proxies to access the internet and you will have to tell R how to get out there. The proxy can be set up manually from within the R console, but it's very annoying when you set a job up to run over the weekend only to come in on Monday morning and find out that the program is waiting for you to type in your user/password and has not even started the run! To avoid this kind of grievance just add another flag to the path:

--vanilla http_proxy=http://proxyname.edu.au:8080/
http_proxy_user=username:password

The 8080 is the port number, it will normally work. Another common port to try is 80. You'll find the proxyname in your browser under settings or ask your IT administrator. Note that you'll leave your username and password exposed as plain text.

You can also run R from a DOS shell (*cmd—Command Prompt*) by simply typing *R* in the command line. Just note that you will either have to be in the directory with R.exe or add it to your path. The default path for the executables are *R-X.XX.X\bin\x64* for the 64-bit build and *R-X.XX.X\bin\i386* for the 32-bit build. Linux and Mac OS add R to the path during installation.

Still in Fig. 1.2 there is the field *Start in*. Here you can change R's default start up directory. When you run R this becomes the active *working directory*, meaning that if no path is given, this is where R will try to read files from or write output to. Not being in the right directory is a common source of frustration when getting started with R. We will discuss this later on but in the meantime it is not a bad idea to set the start up directory to a meaningful path (e.g., *C:\Rprojects*).

Fig. 1.2 Startup properties for R console in Windows. Alongside the path, options for memory size and proxy settings can be added

1.3   Packages and Bioconductor

As previously mentioned, one of the strong points of R is the availability of a huge number of packages to perform different tasks. From the R console, click on the menu *Packages*. The submenu will offer the choice to *load a package* that has already been installed; select the mirror (*Set CRAN mirror*) from which to download packages (choose one close to you—it's usually faster); *Select repositories*, *Install package(s)*, *Update packages*, or *Install package(s) from local zip files*.

Note that the options in the console's menu are simply a *point-and-click* way to execute a command instead of typing it in directly on the console. For example, using the command line in R, the mirror from which to download packages can be set to, e.g., Melbourne in Australia with:

local({r = getOption("repos"); r["CRAN"] = "http://cran.ms.unimelb.edu.au"; options(repos=r)})

A list with the URLs for the mirrors can be found at

http://cran.r-project.org/mirrors.html

Windows packages come as compressed (zipped) files and can be manually downloaded to your machine (not using R) and then installed using the option *Install package(s) from local zip files*. Alternatively, packages can be downloaded and installed straight from R. From the command line you can use *download.packages, update.packages*, and *install.packages*. Be aware of dependencies, packages that need other packages to work properly. The command line option is a safer choice since it will check for dependencies and download/install them for you. Also remember to set the repositories. If say, a package from Bioconductor is needed and the repository is not set you will not be able to download and install the package. In the R console it's simply a matter of selecting the repositories from the popup window. In the command line use: *setRepositories(graphics=F, ind=1:6)*

For analysis of genomic data your one stop shop is Bioconductor. Bioconductor is the repository that holds a large number of packages needed for analysis of genomic data. The link to Bioconductor is *http://www.bioconductor.org/*. There you will find details on how to install packages and groups of packages. There's also lots of information on Bioconductor in general and the various packages. The current version is 2.14.

The easiest way to get started is to download and install the most common packages. In the console type

source("http://bioconductor.org/biocLite.R")
biocLite()

To install a particular package use *biocLite("packagename")*. To download and install everything use *biocLite(groupName="all")*, but be warned that this is a very large installation and definitely not recommended!

Packages are normally installed in the same directory as R, inside a folder called *library*. There's one folder for each package—makes it easy to see what you have installed and to make any changes to a particular package. This all works well if you have unrestricted privileges on the machine. In shared environments you will probably not have write permission to install packages (in Linux, unless you are running R as root this will always be the case) or you might just want to maintain some packages in a different location. For these cases, create a personal library for yourself. That's quite simple from the command line; just add the path to your new *library* when installing packages. For example to download and install a package with all dependencies to the folder *c:/myRlibrary*:

install.packages("PackageName", dependencies=TRUE, lib="c:/myRlibrary")

But R will need to know where to find this library so that it can load packages from it. For this, before trying to load a package, you have to tell R what is the path to your library:

.libPaths("c:/myRlibrary")

And if you do not want to do this each time, add the path to the text file *Rprofile.site* which is found in the *etc* folder in the R directory (you'll need write permission to modify this file, of course!). Another option is to add another flag in the startup properties (as in Fig. 1.2): *R_LIBS_USER="c:\myRlibrary"*.

1.4 R 32-Bit or 64-Bit?

If you are using a 32-bit version of Windows you will have no choice. You can only use R32. If you are running a 64-bit version of Windows you can run R32 or R64. There can be quite a few technical arguments as to which version to run but in practice it mainly comes down to memory. If you have more than 4 GB on your machine you'll probably want to be able to use it, so stick to R64. Some old packages that are not being maintained anymore might only have a 32-bit version, but you will also need an older compatible version of R to run them. The short version is that the world is now 64-bit, 32-bit environments are too constrained for handling genomic data.

1.5 Getting a Handle on R

As soon as you open the console, you are ready to start using R. R can be used as a calculator if you wish. For example at the prompt type in

```
> 43.7*572
```

```
[1] 24996.4
```

and R will give you the answer. Note that by default, input (user commands) is shown in red and output in blue. Just as an aside, pressing *enter* does not necessarily execute the command. For example, you could do the same simple multiplication across two lines

```
> 43.7 *
+ 572
```

```
[1] 24996.4
```

and R will wait until the second value is inserted before multiplying the two numbers. Note that the prompt has changed from > to +, meaning that the console is waiting for additional input. From now on, the + symbol at the start of code line simply means that it is a continuation from the previous line. The > symbol at the start of a line means R is ready to receive the next command.

Of course it's more useful if we can store the results for future use. For this use variables. For example

```
> aa<-43.7
> bb<-572
> cc<-aa*bb
```

now *cc* holds the result. To see the value of *cc* just type

```
> cc
```

```
[1] 24996.4
```

or

```
> print(cc)
```

```
[1] 24996.4
```

The symbol <- means assign or allocate. In R you can allocate values to the left (<-) or to the right (- >). $a <- 43.7$ is the same as $43.7 - >a$. Current versions of R also accept the use of =. So you could also use $a=43.7$. It is recommended to use <- *and* - > in R and considered good practice. I personally don't care! I'd rather use one keystroke than two (plus shift). From now on the = symbol will be used. Just be aware than it is frowned upon!

You can run an entire analysis just typing commands straight into the console. This works well for simple one-off jobs. But most of the time you will be better off by writing a script to perform the tasks you want done and then running it. And by the way, scripts are just a sequence of instructions/commands to be performed. There are several reasons to favor scripts. Many tasks are repetitive—the same script can be used on different datasets. Repeatability—with scripts you always know what you actually did and you can repeat the analysis. Modularity—scripts to perform specific tasks can be combined to perform a larger analysis; over time you will have a nice collection of scripts to choose from. Efficiency—some routines can take a long time to run, it's annoying to stare at a screen for 10 min waiting to write the next line of code.

R Console has a simple editor that can be used to write scripts. But since scripts are simply text files, any text file editor can be used. In Windows a really nice R editor is Tinn-R (Fig. 1.3) which is freely available from

http://sourceforge.net/projects/tinn-r

In Linux, Emacs is a good option; and for Macs, the R editor is reasonable (there is some limited color coding) but TextWrangler is a better choice. Note that you need to install the program and an R syntax highlighting definition file which can be downloaded from the R website (from the main page click on the link *other*; R

Fig. 1.3 Screenshot of Tinn-R with a simple script

GUIs and then *IDE/Script Editors*). On this same page you will also find a large list with many other editors for the three platforms. There are also fully fledged development environments for R (IDEs); Rstudio has become very popular over the last few years. Revolution, a commercial version of R (free for academic use) integrates R with Microsoft's Visual Studio.

To run the script in Fig. 1.3 click on *File* in the R console and then on *Source R code...*, this will open a window to choose the script you want to run. Alternatively use the command *source* and the name of the file on the command line. For example

```
> source("chapter1/Sc1.r")

[1] 24996.4
```

Notice that when using the *Source R code...* option from the menu bar to run a script the console will still output *source("scriptname.r")*. Menu options are just a point and click equivalent to commands.

In the example we used "*chapter1/Sc1.r*". This is because the script was not in the same directory as the current active working directory (if it is you only need to write the name of the script). In this case we did not have to give the full path but only the folder and script name because the script is in a subfolder within the working directory. If the script was in a completely different folder, we would have to use the full path (e.g., *"c:\SomewhereElse\Sc1.r"*). To find out the current working directory type *getwd()* and to change it use *setwd("newpath")*. Or from the menu bar use *File* and *Change dir....*

Throughout the book we will consider that our working directory is one level above the chapter we are working on (for argument sake let's say our working directory is *primer* and there's a separate folder for each chapter).

```
> getwd()
```

```
[1] "C:/primer"
```

One last comment before we get back to the script. Notice that a forward slash is used to specify the path instead of the usual Windows backslash. This is a nice feature in R which helps keep scripts portable across platforms (Linux and Mac use the forward slash for paths). In Windows the R interpreter converts the forward slash into a backslash. It is possible to use the backslash as well but then it has to be double, i.e., two backslashes (\\). We'll discuss the reasons for this later on.

Now back to the script. And we get the exact same result as before using the console. Note: when using scripts the *print* function must be used to explicitly output results to the console, just writing the variable name will not work.

Once the script has run, results (e.g., variables—*aa, bb* and *cc*) are stored in the current R session. It might be useful to keep these results for future use. There are two main options, either save to text files (more details in the next section) or save the whole R session (a.k.a. *workspace*):

```
> save.image("chapter1/myworkspace")
```

and to load a saved workspace:

```
> load("chapter1/myworkspace")
```

Alternatively, workspaces can be saved/loaded using the console's menu (under *File*).

The workspace holds all variables available in the session. To list all variables type

```
> ls()
```

```
[1] "aa" "bb" "cc"
```

It is good practice to remove variables that are no longer necessary. The function for this is *rm*. Let's remove the variables *aa* and *bb*.

```
> rm(aa,bb)
> ls()
```

```
[1] "cc"
```

To remove everything use *rm(list=ls())*, but make sure that you really do not need any of the variables anymore!

Many genomic datasets come as flat (text) files and the structure of these files is not necessarily well documented. Alongside R and a good R editor or IDE, a text editor/viewer capable of opening large data files comes in handy to have a quick look at what's inside the file (e.g., how many lines of comments to remove before the actual data starts). One of my favorites is TextPad

http://www.textpad.com/download/index.html

It's a shareware product costing around US$30.00, but quite worth the price. Unfortunately it will not handle really large datasets (e.g., large fastq files). Some options for large datafiles include:

Large Text File Viewer—http://www.swiftgear.com/ltfviewer/features.html
EmEditor—http://www.emeditor.com/#download
V File Viewer—http://www.fileviewer.com/Download.html

In Linux or Mac OS the terminal programs *more* or *less* can be used. If it's just a matter of having a quick peek at the data, the R *readLines* function does the job well. For example to see the first 3 lines of the script *sc1.r* use

```
> readLines("chapter1/sc1.r",n=3)

[1] "# simple script example"
[2] ""
[3] "aa = 43.7 # assign value"
```

where *n* is the number of lines to read.

1.6 Importing and Manipulating Data

Data is usually stored in binary format, as a flat text file or in a database. Here we will concentrate on text files and in Chap. 3 we will show some examples of how to work with databases. Binary data is a rather complex topic and will not be discussed here (we will however discuss how to save data in binary format to use in R). For the more common sources of data (e.g., Affymetrix CEL files or BAM sequence files) packages to deserialize the binary files are readily available. The reader interested in the topic might find [39] useful.

The most common type of data is in tabular format (rows × columns). The R function to read tabular text files is *read.table*. There are quite a few argument options, for details type *?read.table* (note: documentation for a command can be accessed by *?function_name* or search using *help.search("search string")*).

For example, to import a data file called *snps.txt* into the variable *mydata*:

```
> mydata=read.table("chapter1/snps.txt",
+ header=T,sep="\t")
```

Recall that the plus (+) symbol simply means that the input continues on the next line. And to see the contents of the file:

```
> print(mydata)
```

```
     name allele1 allele2
1    snp1     A       A
2    snp2     A       B
3    snp3     A       A
4    snp4     A       B
5    snp5     B       B
6    snp6     B       B
7    snp7     A       A
8    snp8     A       B
9    snp9     -       -
10   snp10    A       B
11   snp11    B       B
12   snp12    B       B
13   snp13    A       A
14   snp14    A       B
15   snp15    -       -
16   snp16    A       B
17   snp17    B       B
18   snp18    B       B
```

Note here the two main arguments *header,* to define if there is a header in the file (T or TRUE) or not (F or FALSE) and *sep*, the separator (tab in the example). It is always a good idea to check the dimensions of the data, to see if it is what you expected.

```
> dim(mydata)
```

```
[1] 18   3
```

The function *read.table* stores the data as a *data.frame* which is a two-dimensional matrix like structure that can hold different types of data in each column (e.g., numeric, factor, character, etc.). To access the data in a data.frame indexes can be used (e.g., mydata[2,3]) or the column name (e.g., mydata$allele1[3]). For column names the syntax is the name of the data.frame followed by the dollar symbol (*$*) and then the name of the column of interest.

```
> print(mydata[2,3])
```

```
[1] B
Levels: - A B
```

```
> print(mydata$allele1[3])
```

```
[1] A
Levels: - A B
```

The column names can be retrieved with

```
> names(mydata)
```

```
[1] "name"     "allele1" "allele2"
```

To see the top or bottom part of the data use

```
> head(mydata)
```

```
  name allele1 allele2
1 snp1       A       A
2 snp2       A       B
3 snp3       A       A
4 snp4       A       B
5 snp5       B       B
6 snp6       B       B
```

```
> tail(mydata)
```

```
    name allele1 allele2
13 snp13       A       A
14 snp14       A       B
15 snp15       -       -
16 snp16       A       B
17 snp17       B       B
18 snp18       B       B
```

To see a range of data

```
> mydata[1:5,1:2]
```

```
  name allele1
1 snp1       A
2 snp2       A
3 snp3       A
4 snp4       A
5 snp5       B
```

And to see all data in a column use *mydata$name* or *mydata[,1]*.

It is also useful to be able to select a nonconsecutive (and possibly reordered) subset of the data. You can use the concatenate (*c*) function for this. Create a variable that holds the indices you are interested in, for example

```
> indices=c(1,7,10,2,4)
> print(indices)
```

```
[1]  1  7 10  2  4
```

and then show the data for these indices

```
> mydata$name[indices]
```

```
[1] snp1   snp7   snp10 snp2   snp4
18 Levels: snp1 snp10 snp11 snp12 snp13 ... snp9
```

You have probably noticed that after the output there are *levels*. This is an extremely important aspect of R. By default, when a file is read into R and converted into a data.frame, character columns are usually converted to factors (it does make sense, after all R is a statistical programming language). If you do not want R to convert your character columns into factors use *col.Classes="character"* as an additional argument to *read.table*. But note that this will convert all columns to character—not so good if you have numeric data as well. If you have a mix of data types you will need to explicitly define the type of each column (as a vector of types, one for each column). On the other hand, numeric values are imported into R as numbers. Now, why is this important? Because often data has missing values and what the provider of the data used to represent a missing value is anyone's guess! Let's import a file with some animal phenotype measures

```
> pheno=read.table("chapter1/animals.txt",
+    header=T,sep="\t")
> print(pheno)
```

```
          id weight
1    animal1    300
2    animal2    280
3    animal3    350
4    animal4     NA
5    animal5    290
6    animal6    310
7    animal7    300
8    animal8    330
9    animal9    300
10  animal10     NA
11  animal11    280
12  animal12    325
13  animal13    335
14  animal14    305
15  animal15    275
16  animal16    265
17  animal17    415
18  animal18    325
```

We would expect that *id* is a factor and *weight* is numeric. This can be checked by

```
> class(pheno$id)
```

```
[1] "factor"
```

```
> class(pheno$weight)
```

```
[1] "integer"
```

Yes! That worked fine. And looking at the data we can see that there are two missing values that R correctly assigned as missing (*NA*). By default R treats *NA* in the text file as missing. This can be changed using the argument *na.strings*="–" in *read.table* if missing data is represented as "–". Now let's read in the same data but instead of *NA* to represent missing values we have *.

```
> phenowrong=read.table("chapter1/animals2.txt",
+    header=T,sep="\t")
> print(phenowrong)

          id weight
1    animal1    300
2    animal2    280
3    animal3    350
4    animal4      *
5    animal5    290
6    animal6    310
7    animal7    300
8    animal8    330
9    animal9    300
10  animal10      *
11  animal11    280
12  animal12    325
13  animal13    335
14  animal14    305
15  animal15    275
16  animal16    265
17  animal17    415
18  animal18    325

> class(phenowrong$id)

[1] "factor"

> class(phenowrong$weight)

[1] "factor"
```

Ups! Let's fix that.

```
> phenoright=read.table("chapter1/animals2.txt",
+    header=T, sep="\t", na.strings="*")
> print(phenoright)

          id weight
1    animal1    300
2    animal2    280
3    animal3    350
```

```
4    animal4     NA
5    animal5     290
6    animal6     310
7    animal7     300
8    animal8     330
9    animal9     300
10   animal10    NA
11   animal11    280
12   animal12    325
13   animal13    335
14   animal14    305
15   animal15    275
16   animal16    265
17   animal17    415
18   animal18    325
```

```
> class(phenoright$id)
```

```
[1] "factor"
```

```
> class(phenoright$weight)
```

```
[1] "integer"
```

The problem is that too frequently there is no standard for missing data in the same file. Empty spaces, dashes, and words like *none* and *missing* are used all at the same time. This is especially true for phenotypes, covariates, and other information that was collated manually. Rather frequently you will not be able to find all of these just by eyeballing the file. You can *a priori* define the class of each column but if you don't tell R what to treat as missing values you will get an error when reading in the file. So, always check the classes in the data.frame to make sure that they are what you expected.

Now, how to fix this? The easiest way is to convert the column from *factor* to *numeric* format using *as.numeric*. But there is a catch here. First convert from *factor* to *character* and then to *numeric*! Forgetting to do this is one of the main sources of problems people come across in R! Let's exemplify.

```
> as.numeric(phenoright$weight)
```

```
 [1]  300 280 350   NA 290 310 300 330 300   NA 280
[12]  325 335 305 275 265 415 325
```

```
> as.numeric(phenowrong$weight)
```

```
 [1]   6  4 12  1  5  8  6 10  6  1  4  9 11  7  3
[16]   2 13  9
```

In the first case the conversion did what was expected (ok—it was already numeric anyhow, but you get the point). In the second case, you also got numbers—but they are definitely wrong. What are these numbers? They are the indices of the factors!

If extreme care is not taken you might end up doing calculations using indices and not the values of your measurements. So, one way to convert is

```
> as.numeric(as.character(phenowrong$weight))

 [1] 300 280 350   NA 290 310 300 330 300   NA 280
[12] 325 335 305 275 265 415 325
```

That does the trick—you do get a warning that *NAs* were introduced, but that's what you wanted anyhow. And to fix your data.frame

```
> phenowrong$weight=
+ as.numeric(as.character(phenowrong$weight))
> class(phenowrong$weight)

[1] "numeric"
```

The values in *weight* are integers, so instead of *as.numeric* you could also have used *as.integer*.

Often we want to combine different datasets into a single data.frame. A simple way of doing this is using *cbind* to bind columns or *rbind* for rows. Note that the number of rows must match in the first case and the number of columns in the second. For example to make a single data.frame from *mydata* and *pheno*

```
> alldata=cbind(mydata,pheno)
> print(head(alldata))

  name allele1 allele2      id weight
1 snp1       A       A animal1    300
2 snp2       A       B animal2    280
3 snp3       A       A animal3    350
4 snp4       A       B animal4     NA
5 snp5       B       B animal5    290
6 snp6       B       B animal6    310
```

If we look at *alldata* it is quite obvious that the columns *name* and *id* are equivalent (another classic problem—different identifiers used for the data and they don't even have the same meaning). We will see some examples of this and how to merge datasets later on. For the time being let's just reorder the data and drop the redundant identifier *name* which is at best misleading (the data is for a single SNP genotyped on many samples and not many SNP for a single sample).

```
> alldata=alldata[,-1] # use minus (-) to drop a column
> # now reorder the data.frame the way we want it
> alldata=alldata[,c(3,4,1,2)]
> print(head(alldata))

       id weight allele1 allele2
1 animal1    300       A       A
2 animal2    280       A       B
```

```
3 animal3      350        A        A
4 animal4       NA        A        B
5 animal5      290        B        B
6 animal6      310        B        B
```

The minus sign followed by an index can be used to delete a column. This is quite a useful feature of R and is not limited to a single column. Before we used a variable called *indices* to subset data (*mydata$name[indices]*), if we used *–indices* instead we would get all data except for the values in rows 1, 7, 10, 2 and 4. To reorder the data.frame we use *c* and the order in which we want the columns. We might also want to sort the data. Let's sort by *weight*.

```
> alldata=alldata[order(alldata$weight,decreasing=T),]
> print(alldata)

          id weight allele1 allele2
17 animal17    415        B        B
3   animal3    350        A        A
13 animal13    335        A        A
8   animal8    330        A        B
12 animal12    325        B        B
18 animal18    325        B        B
6   animal6    310        B        B
14 animal14    305        A        B
1   animal1    300        A        A
7   animal7    300        A        A
9   animal9    300        -        -
5   animal5    290        B        B
2   animal2    280        A        B
11 animal11    280        B        B
15 animal15    275        -        -
16 animal16    265        A        B
4   animal4     NA        A        B
10 animal10     NA        A        B
```

Now our data.frame is sorted by column *weight* in decreasing order (if you want the data sorted in increasing order use *decreasing=F*).

1.7   Plots and Descriptive Statistics

Once we have read the data file into R, checked that the dimensions are what would be expected and converted the columns into the correct format, we are ready to start working with the data. Before doing any fancy analysis it is always a good idea to get a *feel* for the data with some basic plots and descriptive statistics. A very handy function in R is *summary*.

```
> summary(alldata)

          id                 weight         allele1 allele2
 animal1 : 1     Min.    :265.0      -: 2      -: 2
 animal10: 1     1st Qu.:287.5      A:10      A: 4
 animal11: 1     Median :302.5      B: 6      B:12
 animal12: 1     Mean    :311.6
 animal13: 1     3rd Qu.:326.2
 animal14: 1     Max.    :415.0
 (Other) :12     NA's    :  2.0
```

The *summary* function will give you counts for *factors* and measures such as mean, median, and others (as above) for numeric columns (*weight* in our example). The results for *id* are clearly not relevant but for *allele1* and *allele2* you can immediately see how many of each allele there are and also the number of missing genotypes (by the way, the symbol for missing here was "–"). You can also get each of these statistics independently

```
> min(alldata$weight,na.rm=T)

[1] 265

> max(alldata$weight,na.rm=T)

[1] 415

> mean(alldata$weight,na.rm=T)

[1] 311.5625

> median(alldata$weight,na.rm=T)

[1] 302.5

> quantile(alldata$weight,0.25,na.rm=T)

  25%
287.5

> quantile(alldata$weight,0.75,na.rm=T)

   75%
326.25
```

Notice the use of *na.rm=T* which means remove missing values. If you forget to add this you will get *NA* as an answer.

```
> mean(alldata$weight)

[1] NA
```

Some other useful summaries are standard deviation (*sd*) and variance (*var*)

```
> sd(alldata$weight,na.rm=T)

[1]  36.36476

> var(alldata$weight,na.rm=T)

[1]  1322.396
```

As the old adage goes, a picture is worth a thousand words—and that's very true for summarizing data (and especially genomic data, as we will see in the next chapters). Let's plot the counts for the alleles in both columns using a *barplot* (Fig. 1.4).

```
> barplot(c(summary(alldata$allele1),
+    summary(alldata$allele2)),
+ main="Allele counts",col=c(1,2,3,1,2,3))
```

Notice how we are starting to chain R functions together. To create the barplot, we first used the *summary* function to obtain the counts (number of "–", "*A*", and "*B*") for each of the two allele columns, then we concatenated the counts into a vector using *c*. Finally the, in this case, six counts were provided to the *barplot*

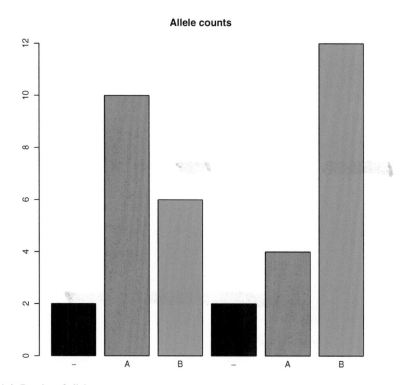

Fig. 1.4 Barplot of allele counts

function to create the image. The argument *main* is used for the graph's title and *col* is for colors. To see which colors are available in R type *colors()* or *palette()* for the active colors. You can use the actual name of the color or a numeric value as in the example. The numeric value refers to the index of the color returned by *palette()*.

```
> palette()
[1] "black"   "red"     "green3"  "blue"
[5] "cyan"    "magenta" "yellow"  "gray"
```

We could also pool all alleles together and then look at the overall counts (Fig. 1.5).

```
> pooled=c(as.character(alldata$allele1),
+   as.character(alldata$allele2))
> pooled=summary(factor(pooled))
> barplot(pooled,main="Pooled allele counts",col=
  c(1,2,3))
> legend("topleft",c("-","A","B"),fil=1:3)
```

Now we split the plotting into two parts. First we made a variable with the pooled data (very creatively called *pooled*) and then used this variable for plotting. We also

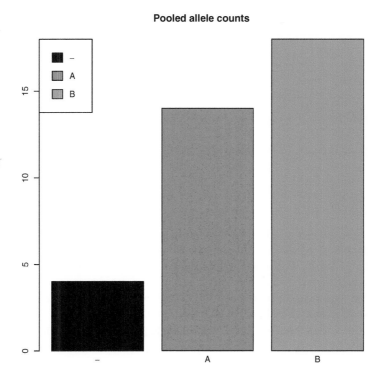

Fig. 1.5 Barplot of pooled allele counts

added a legend to the plot. The first argument is the position where we want the legend, then the captions and finally the color coding (*fil*). Notice the workaround to merge the two allele sets. Again we run into trouble because of factors in R. You cannot simply concatenate two sets of data of class *factor*.

```
> head(c(alldata$allele1,alldata$allele2))
```

```
[1] 3 2 2 2 3 3
```

```
> head(c(as.character(alldata$allele1),
+    as.character(alldata$allele2)))
```

```
[1] "B" "A" "A" "A" "B" "B"
```

See? Again you end up with the indices! And if you try to summarize the indices, R will treat them as numeric and give you

```
> summary(c(alldata$allele1,alldata$allele2))
```

```
   Min. 1st Qu.  Median    Mean 3rd Qu.    Max.
  1.000   2.000   2.500   2.389   3.000   3.000
```

Which is definitely not what you want. After converting the factors into *character* we have to transform them back into *factor*! Otherwise R will treat the data as *character* and your summary will look as follows:

```
> summary(c(as.character(alldata$allele1),
+    as.character(alldata$allele2)))
```

```
   Length     Class      Mode
       36 character character
```

So, the take home message is: *always pay attention to what is happening with factors!*

Another common plot is the *boxplot*. Of course it will only work with numeric data, so let's use *weight* (Fig. 1.6).

```
> boxplot(alldata$weight,col="blue",
+    main="Boxplot of weights for all animals")
```

The hinges show the extreme values excluding outliers, the box borders are the upper and lower quartiles, and the line in the box shows the median. Outliers are flagged as circles. In the example we have one outlier—which should be investigated before further analysis (i.e., is it a real value or a data error that should be removed).

Now we have a visual representation of the data we get using *summary*. In general we are more interested in looking at the distribution of one variable in relation to the different classes the observed values belong to. This is very easy in R. Let's see what *weight* looks like if we split the data using the levels in *allele1* (Fig. 1.7).

```
> boxplot(alldata$weight~alldata$allele1, col=2:4,
+    main="Boxplot of weights by allele class")
```

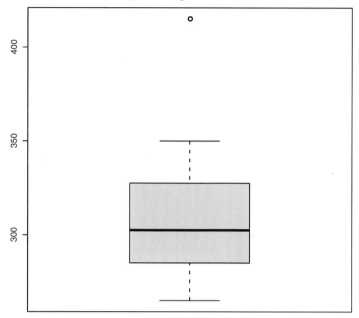

Fig. 1.6 Boxplot of weights for all animals

The tilde symbol (˜) means "modeled by". Here we have *weight* modeled by *allele1*. In Chap. 2 we will return to boxplots in more meaningful scenarios.

Another good way of visualizing data is with density plots (Fig. 1.8).

```
> plot(density(alldata$weight,na.rm=T),
+    main="Density plot of weights",col="blue")
```

It is immediately obvious that the data is reasonably normal except for the outlier—which adds the extra hump to the density plot. Note again the use of *na.rm=T* or you will get an error. Here we first calculated the density and then plotted it. The function *plot* is the most generic plotting function in R with many specific methods for different objects. To exemplify, let's plot the sorted values of *weight* and then connect each point with a line through the points (Fig. 1.9).

```
> plot(sort(alldata$weight),col="blue",
+    main="Sorted weights",xlab="animal",ylab="weight")
> lines(sort(alldata$weight),col="red")
```

This may at first look a bit complicated, but all we are doing is sorting the data with the function *sort* and then plotting the points. The arguments *xlab* and *ylab* are the labels for the *x* and *y* coordinates of the plot. And then we use *lines* to add a line through our points.

Boxplot of weights by allele class

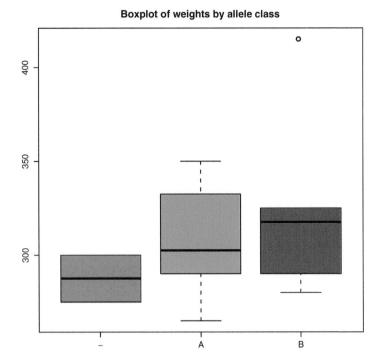

Fig. 1.7 Boxplot of weights for all animals separated by allele class

The graphical capabilities or R are extensive. We have just skimmed over the surface of some useful options. We will look at other graphs as we progress the topics. Try having a look at some other options such as *pie* or *hist*. Excellent reference books for graphics in R are [19, 70, 78] and [61].

1.8 Saving Results

We have already seen how to save the entire workspace. But most of the time all you want to do is save the relevant results—for example the *alldata* modified data.frame that we made. For this you can use the function *write.table*.

```
> write.table(alldata, "chapter1/alldata.txt",
+    quote=F, row.names=F, sep="\t")
```

This will save the data.frame *alldata* as a text file *alldata.txt* in the directory *chapter1* (sideline: R recognizes \ as an escape symbol—as used, e.g., for tab (\t)— to get an actual backslash in R you must use a double backslash: \\). By default *write.table* saves values between quotes, to remove them use the argument *quote=F*. You also probably will not want the row names (at least not in this case), use

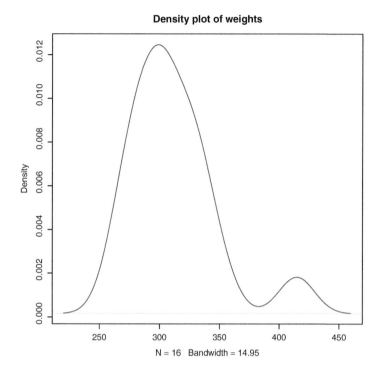

Fig. 1.8 Density plot of weights for all animals

row.names=F to remove them. If you don't want a header use *col.names=F*. To save a file that can be opened straight into a spreadsheet such as Excel, you could call the file *alldata.csv* and use *sep= ","* as a separator.

A down and dirty way to save results is to redirect the console output to a text file using *sink*.

```
> sink("chapter1/out.txt")
> alldata
> sink()
```

This will redirect the console output to a file called *out.txt*. Remember to close the pipe when you are finished using *sink()*. There are many other ways of saving data in R. Have a look at the help files for *save, write, writeLines* among others.

You will also want to save plots for reports or publications. This can be done manually: go to the plot window, click on *file* and then save the image. Programmatically, use the function *dev.print*. To save the density plot as a *pdf*

```
> plot(density(alldata$weight,na.rm=T),
+    main="Density plot of weights",col="blue")
> dev.print(file="chapter1/density.pdf",
+    device=pdf,width=8,height=8)
```

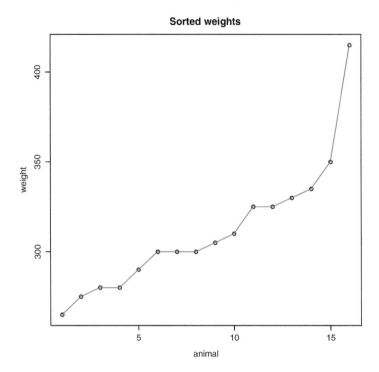

Fig. 1.9 Sorted plot of weights for all animals

The arguments are the file name, format (e.g., *pdf* or *png*) and dimensions—height and width. For *pdf* the dimensions are in inches but for *png* they must be given in pixels (e.g., *width = 1,024, height = 768*).

1.9   Some Help on Help

I have not been overly kind to R's help files so far. I even mentioned that the R help is rather unhelpful. But as you come to terms with R you will get to know the names of the functions and then it becomes very useful.

Apart from the *help.search("function_name")* and *?function_name* discussed before, you can also use *help.start()* or click on *Help* and then *HTML help* to open a browser window on the main help page (HTML help is the default option during installation). Just note that the last few releases of R need an HTTP server to be running for the search engine to work. This is normally not a problem if you launch help from within an R session, but search will not work out of an R session (unless the server is up and running). Further, as soon as you close the R session your help session goes with it (the current page will remain open but the links will not work anymore). And a last hint that is sometimes useful is to type the first few letters of a function followed by a double *tab*. This will show some of the available functions that start with those letters. By the way, a single *tab* can be used for auto-completion of a function's name, but it has limited functionality since it will only do something after what has been typed in matches a single function.

1.10   Where to Go from Here?

This ends our rather quick tour of R. I have tried to cover some of the most basic procedures that are regularly used: importing, summarizing, visualizing and saving data. But as I mentioned before, there are hundreds of packages—so far we have only made use of the basic R installation—and there are also dozens of books covering a plethora of topics. This chapter ends by mentioning a few resources that might help to get you going down the R path (we'll leave the *bio* references for the other chapters). This list is in no way exhaustive and admittedly rather biased on my own preferences.

- When you install R you have the option of installing *pdf* manuals as well. Make sure you install these. There's an excellent *introduction to R* which will cover what we discussed here in much greater detail. The reference manual includes all functions from the base installation—a search in this manual can be more fruitful than the help search engine. There are also a few more advanced manuals worthwhile reading.
- From the R website (http://cran.r-project.org/) you can download whole books which are freely available and also some handy reference cards. It is also worthwhile reading the FAQs.
- For a *user's manual* style introduction to R (i.e., R as software), you should start with the short text by Zuur et al. [126]. It was also written with life scientists in mind.
- Crawley [20], Dalgaard [22], and Verzani [114] are good texts for introductory statistics and R. The more formal and less introductory Cohen and Cohen [17] is also well worthwhile reading.
- Crawley's *The R Book* [21] is huge but it is a really good text that covers most types of analyses (albeit at an introductory level). This one and Adler's [1] *R in a Nutshell* complement each other very well.
- For linear models try Faraway [31], the example driven book by Everitt and Hothorn [29] or the excellent regression book by Sheather [100]. More advanced models are covered in Faraway [32], the classic Pinheiro and Bates [86] book on mixed models which is based on S but translates well into R, and Dobson et al.'s [24] generalized linear models text.
- For data manipulation Spector's [105] text covers most common needs.
- R has excellent plotting capabilities. A good coverage of lattice plots is given by Sarkar [96] and for more general plots have a look at Maindonald and Braun [70]

and also Cook and Swayne [19], Murrell [78] (the second edition should be out by the time you are reading this text), and Keen [61].

- Bayesians will like Albert [4] and the more comprehensive Carlin and Louis [14]. Probability simulation and Gibbs sampling is discussed by Suess and Trumbo [107] and is a good companion to the text on Monte Carlo methods by Robert and Casella [15].
- Probability is covered extensively in Ugarte et al. [112] and a lighter read (at least in physical dimensions!) is Baclawski [9].
- A bit more programmatic in nature, the books by Rizzo [94] or Jones et al. [60] give some good foundations of the underlying algorithms used for statistical analyses and on the usage of R as a programming language. And if programming is the order of the day, you should not do without Chamber's [16] *Software for Data Analysis*.

Chapter 2
Simple Marker Association Tests

In this chapter we will briefly overview genetic markers and their use in association studies. Then we will discuss step by step how to import data from a small genomic project, check the data for inconsistencies, fit and evaluate different statistical models and interpret the results.

2.1 Introduction to Markers

Genetic markers can be described as a measurable variation (*polymorphism*) at the DNA level. They are chromosome regions where differences in nucleotide sequences occur between individuals of the same species. Various types of markers have been used to detect DNA variability, the most common being:

- Restriction fragment length polymorphism (RFLP)
- Random amplification of polymorphic DNA (RAPD)
- Amplified fragment length polymorphism (AFLP)
- Single stranded conformation polymorphism (SSCP)
- Copy number variation (CNV)
- Microsatellites
- Single nucleotide polymorphism (SNP)

The last two, microsatellites and SNP are currently the most commonly used markers—we will focus on these herein. This might be somewhat academic but note that there is a difference between a genetic marker and DNA variation. A marker is more related to the technique employed to measure the variation whilst DNA variation is obviously independent of our ability to measure it or not. When we talk

Electronic supplementary material The online version of this chapter (doi: 10.1007/ 978-3-319-14475-7_2) contains supplementary material, which is available to authorized users.

© Springer International Publishing Switzerland 2015
C. Gondro, *Primer to Analysis of Genomic Data Using R*, Use R!,
DOI 10.1007/978-3-319-14475-7_2

about DNA variation *per se* we should think more in terms of indels—insertions or deletions of various lengths; VNTRs—variable number of tandem repeats, which are multiple copies of the same sequence of DNA, usually simple non-coding repeats (e.g., *AC* repeated a certain number of times); and SNP—single nucleotide polymorphisms which are a single nucleotide mutation (base substitution). SNP occur in very large numbers and are widely spread across the genome in both coding and non-coding regions, which makes them ideal for association studies.

The objective of an association study is to link a specific DNA region (and its underlying variation) with a trait of interest. For example in human research, many studies have been undertaken to identify genomic regions which confer greater resistance/susceptibility to diseases such as cancer (an accessible overview of association studies in cancer is given by Patel et al. [85]), and ultimately, within these regions, identify the variants that lead to expression of the trait. Knowledge of these *functional* variants improves our understanding of disease mechanisms, suggests targets for novel therapeutics and can be useful for risk prediction and prevention. For the latter, in a very naive way, this would mean that in the future a new individual could be genotyped and if he/she has SNP variant *A* at location *23445635* on chromosome *3*, then this individual has a 26 % probability of developing Alzheimer. In livestock the objective is to find associations that can explain part of the variance observed in a production trait. For example one of the variants found in microsatellite marker *M* is associated with an increase in marbling of *x*%. This of course has implications for breeding decisions and can help improve the accuracy of breeding values. Identification of regions associated with a trait can also help select candidate regions for gene discovery—that is, identify the actual genes ("*the real biology*") and the functional mutations that confer the trait differences (e.g., myostatin for double muscling). Note that an association does not imply causality. A marker can be associated with a trait but it is not necessarily in any way related to it. Recall that a chromosome is a physical structure of nucleotides connected to each other which only gets broken up by recombination events. Hence we can think of DNA sequences being transmitted from parent to offspring in *chunks*. The closer the marker is to the actual causative gene, the greater the chance that the same marker allele will be transmitted to the offspring with the same causative allelic variant of the gene—the marker and the gene are in *linkage*, on the same DNA *chunk*, or more formally *haplotype*. A haplotype is er, hmm, a *chunk* of DNA which has a combination of alleles that would not be observed by chance if each one was segregating independently. This brings us to the concept of *linkage disequilibrium* (LD): a non-random association of alleles. Pay some attention to the difference between linkage and linkage disequilibrium. The first is more related to the actual physical connection of nucleotides on a DNA strand. The second is more of a statistical concept, we can detect LD with a statistical test of significance but it is not necessarily telling us anything about the actual physical proximity of what we are testing (although reasonable inferences can be made).

In short, the terminology is rather telling. Genetic markers mark, they don't necessarily cause and association tests identify connections: what we can measure (marker) is to some extent related to what we want to know (e.g., the true causative gene).

Just a couple of additional points we should keep in mind: the perfect marker is a causative marker, as for example the anecdotal SNP we used to exemplify (partial) susceptibility to Alzheimer. But if the marker is not causal, the further away it is from our causative gene the greater the chance that a recombination event will break up the linkage between marker and gene. To illustrate, in our also anecdotal marbling example if a recombination occurred between marker *M* and the *true marbling gene* in the vicinity of the marker, the *favorable* variant would then no longer be useful to track the causal variant.

2.1.1 Microsatellites

Microsatellites are a type of VNTR. If the length of the repeating unit is less than five base pairs, the VNTR is referred to as a microsatellite. If the length of the repeating unit is greater than five base pairs the VNTR is called a minisatellite.

Variation in microsatellites is measured through the number of repeats that an individual has of the repeating unit. For example, back to our *AC* repeating unit, one individual might have ten repeats of *CA* while another has 7. It is immediately evident that these regions will have different sizes and, if we manage to quantify them, we can use these measures to distinguish individuals. We can get a handle on microsatellites using PCR—polymerase chain reaction—methods. Briefly, unique primers that flank the repeating sequence are used to amplify the region and then the amplified product is separated by size using a more manual gel electrophoresis or a high throughput capillary system. We can refer to each of these different sizes as an *allele*. Note that diploid organisms will have two of each chromosome and consequently can have two different microsatellite alleles (*heterozygote*) or two of the same (*homozygote*). Just as a curio, we used the *AC* example because it is the most common repeat in humans—dinucleotide repeats are the most frequent and 50 % of these are *AC*.

As we discussed above marker alleles are not necessarily causative of the trait we are interested in. In the case of microsatellite markers it is almost a given that they are not causative and that they are in non-coding regions (it can't do much good to a functional gene to have variable numbers of repeating sequences inside it). Another point to mention with microsatellites is that they can have many different alleles (SNP will generally only have two), between 10 and 30 is quite common. This can be a blessing and a curse at the same time. A single microsatellite marker is much more informative (many alleles) than one SNP marker (mostly only two alleles). On the down side, they are much sparser in the genome (microsatellites are in the thousands while SNP are in the millions) with much lower LD between marker and causal variant.

We will defer discussing SNP until the next chapter. Here we will limit the discussion to analysis of microsatellites under the scenario that we have identified a candidate region from say, a genome wide mapping project (e.g., [71]), we have genotyped some microsatellite markers in this region and then we test this fine mapped region for association. Microsatellites have been widely used in QTL mapping projects and there is an extensive literature in the field in relation to experimental design and appropriate analytical approaches. Keep in mind that the example we will work on in this chapter is much more akin to a genome wide association study using SNP (but with just a few markers instead of thousands) than a traditional QTL mapping study with sparse markers. Excellent texts on linkage analysis and QTL mapping are [120] and [102].

But before we delve into analyses using R let's have a quick look at the two most common designs for association studies: case–control and family-based, as well as the types of traits—discrete (often binary) and continuous or quantitative.

2.2 Case–Control and Family-Based Association Studies

Case–control is a classic approach used in epidemiological studies. The *cases* are individuals who show the disease (in epidemiological or medical research) or express the trait/condition under study. The *control* is made up of individuals that do not express the disease or trait. The objective is to compare the frequency of alleles and/or genotypes of genetic markers between the two groups to determine if any of the genotyped polymorphisms are associated with the disease/trait and what is the probability that this polymorphism will increase/reduce the chance of the individual (and other individuals in the population with the same polymorphism) expressing the trait or being afflicted by the disease. A point to note is that genotypic and haplotypic frequencies can vary quite significantly across populations. The experiment has to be designed is such a way as to minimize or be able to account for the effect of *population stratification* which can confound your results. For example, Korean Hanwoo cattle are highly marbled while Brahman cattle have low intramuscular fat. If we set up an experiment to contrast Hanwoo and Brahman, marbling (the trait) would be completely confounded with breed effect. In effect, breed differences are probably one of the higher order associations that can be easily picked up by differences in allelic frequencies, as shown in the nice paper by Hayes et al. [53]. So, to set up a case–control study try to minimize population stratification by selecting cases and controls with similar genetic backgrounds or at least select samples in a way that will allow separating population structure from the trait (somewhat akin to a *block design*).

Family-based association designs are, as the name suggests, based on family relationships between individuals in the test. It has the advantage over case–control studies in the ability to reduce confounding effects of population stratification. The general principle is to compare siblings which have the same genetic background and test if either one of the allelic variants of the parent (of course the parent has to be heterozygous for the marker) is significantly associated with the disease or trait. As a simple example, take a sire that expresses a trait of interest (e.g., very high resistance to tick infestation), is heterozygous for a *suspect* marker and mate this sire with many homozygous females that do not express the trait, preferably they will be homozygous for a third allele, different from the sire's alleles. Then collect phenotypes of the offspring, genotype them for the marker, and test if there is an association between the allele they inherited from the sire and their level of resistance to tick infestation. This is a simple example of a half sib design. There are also full sib designs and other more complex ones. In humans, for obvious reasons, there's limited input of the researcher into the mating structure, so it's common to work with many small families with a heterozygous parent and one normal/one affected offspring—the family trio. In livestock there are fewer restrictions on setting up the experimental design and then questions relating to how many families and family sizes become extremely relevant. We will not go into any in-depth details in this text, but power and sample size are critical considerations in genomic studies and should be given proper attention before embarking on a large scale experiment. A paper by Spencer et al. addresses the main points of design within the scope of genome-wide association studies [106]; see also the book chapter of Roderick Ball with nice R examples [10]. Some other R packages that can assist with power tests are listed at the end of the chapter.

2.3   Discrete and Quantitative Traits

So far we have focused on the markers. Let's now give some consideration to the type of traits we could be interested in. In humans most disease traits are treated as discrete and generally binary, i.e., you either have the disease or you don't. A livestock example is marbling score, where animals are assigned a value between 1 (very low marbling) and 12 (very high marbling). The analysis of a case–control study for a discrete trait with only a few levels and a single marker can be as simple as a χ-square contingency test.

On the other hand, we also have continuous traits such as height or weight which are quantitative traits. These traits usually follow a normal distribution. Here also there's no need for anything too fancy. Again in a case–control study, if you just want to test the difference between two alleles, a simple t-test might do the trick. And for a few alleles an Anova is a good starting point. Things get a bit more complicated once you have to start taking into account population stratification and fixed effects that impacted on your observations. But we will get back to this further on.

With population-based studies the pedigree clearly plays a main role in the analysis. How to tackle the issue will depend on the structure of the relationships. In livestock it is not unusual to have half sib designs—a few (10–20) sires mated to randomly selected females from the population and each sire has again 10–20 offspring (that's what we will be working with in our example further on). An easy way to analyze this data is to test for association individually within sire families and compare the results across families. It's also worthwhile to pool the data into a single combined analysis.

Just before we move on, the message here is that it is crucial that you understand what type of trait you are working on and the data structure. The distribution of the trait is paramount to making a decision of an adequate model for your data.

2.4   Additive, Dominant, and Recessive Models

The main models normally used for a single marker are the *additive*, *dominant*, or *recessive* model. If we go back to our basic genetics classes we will recall that we were endlessly drilled about the expected phenotypic ratios given that this allele was dominant over that one or what were the expected phenotypic proportions if this allele is recessive in relation to that other one and etc. Genetics has not changed, so it's still the same old story. When we test for association we have to consider if, once an individual has certain allele, whatever the other allele is, the phenotype will be the same—a dominant model, $AA = Ax$ and x can be any other allele. Alternatively, we can have the inverse. If an individual has allele a (just to stick with the classic notation), its phenotype will be determined entirely by the other allele (unless of course it's the same a)—that's the recessive model. The third model—additive—considers that each allele individually contributes to the phenotype, thus an individual Aa would have a phenotype that is the simple sum of the estimated individual values of alleles A and a.

A more generic framework is to use a *genotypic* model. Here instead of estimating allelic effects and the interaction between alleles, we estimate allelic pair combinations (genotypes). This has the advantage that the heterozygote does not need to be the sum of two allelic effects, it can take any intermediate value, it can be beyond the values of the homozygotes on either side—overdominance, and we can also quite easily detect dominance or recessive effects (e.g., no difference between AA and Aa). This rather trivial example with only two alleles and three genotypes (but it becomes relevant with SNP) suggests that the genotypic model is a good *one size fits all* approach; however if we have many markers there might not be enough data to estimate the genotypic effects, e.g., for each marker there are three or four alleles that the offspring inherited from the sires and a whole lot of other alleles from the various dams.

If we go beyond single markers, epistasis comes into play. This is a hot research topic nowadays, we will not discuss it here but always keep in mind that life is complex.

2.5   A Worked Out Example

But this text is meant to be applied, let's get back to the main point: how to do a marker association test using R.

First note that there will be many intermediate steps before we perform the actual association test—they involve understanding the structure of the data and performing formatting and clean up steps for the analysis. Frankly, this can be rather boring but it is important to get a handle on how to perform these preprocessing steps in practice since they will be routinely required in real world applications.

Suggestion: as you work your way through the example, write the code in an editor (use e.g. Tinn-R). Some parts are quite long and you will not want to retype everything if you make a mistake. Also, if something goes very wrong, it will be quite easy to rerun from scratch if you save all the steps.

We will use a simulated dataset (but in a realistic format). Here's our scenario: from previous mapping experiments a certain genomic region was identified as a potential QTL region (by the way, a QTL is a quantitative trait locus—a genomic region that explains phenotypic differences due to genetic polymorphisms) associated with weight in cattle. Even though there is some confidence that there really is a QTL in this region, its exact location is not well defined (the confidence interval is 10 cM). Researchers looked at the actual known genes in this region (details in Chap. 6) and based on the functional knowledge of these genes decided that five could potentially be involved in the trait. Five microsatellite markers that we expect to be in full linkage with each of these genes were identified and selected for the project. The researchers then set up a half sib experimental design with 10 heterozygote sires and each sire had 40 offspring (perfect numbers that only simulations will allow!) with randomly selected females from a population with a similar genetic background. The sires and the offspring were genotyped for all five markers and phenotypic measures were recorded (the females were neither genotyped nor measured). We received this dataset and our task is to test for association between these markers and the phenotypes. Hopefully we will find a marker that is associated with our trait and we will be able to get closer to pinpointing the actual causative gene! Let's get started...

It's a very small dataset, *10 sires ×40 offspring = 400 records*, plus the genotypes/phenotypes from the sires themselves. We received two data files: one with the progeny and another with the sires. Our files are called *progdata.txt* and *siredata.txt*. The first step is to open the data in a text editor and look at how it is formatted (Fig. 2.1).

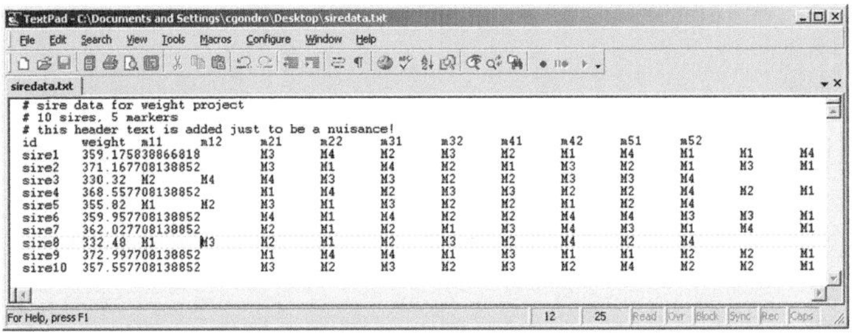

Fig. 2.1 Screenshot of raw sire data

We immediately see that there are three lines of information that are not part of the data itself; *progdata.txt* also has three lines of *noise*. We start by importing the data into R and checking if everything looks fine.

```
> sires=read.table("chapter2/siredata.txt",
+       header=T,sep="\t",skip=3)
> prog=read.table("chapter2/progdata.txt",
+       header=T,sep="\t",skip=3)
> dim(sires)

[1] 10 12

> dim(prog)

[1] 400  14

> print(sires)

        id weight m11 m12 m21 m22 m31 m32 m41 m42 m51 m52
1    sire1 334.14  M2  M1  M3  M2  M3  M4  M4  M2  M4  M2
2    sire2 364.81  M3  M2  M3  M2  M2  M3  M2  M4  M3  M1
3    sire3 383.95  M2  M4  M2  M4  M3  M2  M3  M4  M1  M4
4    sire4 349.88  M2  M1  M1  M2  M4  M3  M2  M1  M4  M3
5    sire5 357.87  M1  M3  M2  M1  M3  M1  M3  M4  M3  M2
6    sire6 364.87  M3  M2  M1  M3  M4  M2  M3  M1  M1  M3
7    sire7 361.36  M4  M1  M4  M1  M2  M1  M1  M2  M2  M4
8    sire8 357.56  M3  M4  M2  M4  M4  M1  M2  M4  M1  M3
9    sire9 333.49  M1  M2  M4  M3  M3  M2  M1  M2  M3  M2
10  sire10 360.92  M2  M3  M3  M1  M3  M2  M2  M3  M3  M2

> head(prog)

   id   sire sex weight m11 m12 m21 m22 m31 m32 m41 m42
1 id1 sire1   F 293.61  M5  M6  M2  M5  M3  M6  M2  M5
2 id2 sire1   M 335.43  M1  M4  M3  M1  M4  M5  M4  M3
3 id3 sire1   M 340.09  M2  M3  M2  M6  M3  M1  M4  M3
4 id4 sire1   M 343.08  M2  M3  M2  M1  M4  M6  M2  M3
5 id5 sire1   F 287.08  M1  M3  M2  M4  M4  M6  M4  M5
6 id6 sire1   F 302.17  M2  M2  M2  M5  M3  M5  M2  M4
  m51 m52
1  M4  M4
2  M2  M2
3  M4  M3
4  M4  M5
5  M2  M3
6  M4  M1
```

Everything looks ok. We have the correct numbers in both files. Notice the use of *skip* as an argument to jump the first three lines.

So, what have we got on data? In *prog* we have the animal id, should be a unique identifier for each animal, we have sex, weight—our trait of interest and the marker information, one column for each marker and each allele, apparently the first number refers to the marker and the second to the allele (i.e., *m11* is marker one, allele one). To build genotypes we will have to pool together two columns, but that's later on. Let's summarize the data and see if there's anything amiss going on.

```
> summary(sires)
```

id		weight		m11	m12	m21	m22
sire1	:1	Min.	:333.5	M1:2	M1:3	M1:2	M1:3
sire10	:1	1st Qu.:351.8		M2:4	M2:3	M2:3	M2:3
sire2	:1	Median :359.4		M3:3	M3:2	M3:3	M3:2
sire3	:1	Mean	:356.9	M4:1	M4:2	M4:2	M4:2
sire4	:1	3rd Qu.:363.9					
sire5	:1	Max.	:383.9				
(Other):4							

m31	m32	m41	m42	m51	m52
M2:2	M1:3	M1:2	M1:2	M1:3	M1:1
M3:5	M2:4	M2:4	M2:3	M2:1	M2:4
M4:3	M3:2	M3:3	M3:1	M3:4	M3:3
	M4:1	M4:1	M4:4	M4:2	M4:2

```
> summary(prog)
```

id		sire		sex	weight	
id1	: 1	sire1	: 40	F:200	304.06 :	3
id10	: 1	sire10	: 40	M:200	299.28 :	2
id100	: 1	sire2	: 40		307.21 :	2
id101	: 1	sire3	: 40		311.7 :	2
id102	: 1	sire4	: 40		313.62 :	2
id103	: 1	sire5	: 40		313.66 :	2
(Other):394		(Other):160			(Other):387	

m11	m12	m21	m22	m31	m32
- : 1	- : 1	- : 1	- : 1	- : 1	- : 1
M1: 98	M1:70	M1: 97	M1:64	M1: 55	M1:66
M2:139	M2:68	M2:130	M2:63	M2:115	M2:69
M3:101	M3:67	M3: 93	M3:79	M3:134	M3:72
M4: 60	M4:62	M4: 79	M4:61	M4: 95	M4:61
M5: 1	M5:74		M5:56		M5:69
	M6:58		M6:76		M6:62

m41	m42	m51	m52
- : 1	- : 1	- : 1	- : 1
M1: 78	M1:57	M1: 83	M1:58
M2:142	M2:68	M2: 94	M2:64
M3: 82	M3:78	M3:135	M3:64

```
M4:  97      M4:75      M4:  87      M4:74
             M5:68                   M5:76
             M6:53                   M6:63
```

The *sires* data seems fine. For each marker there are four alleles (once we pool the data from *m_1* and *m_2* together). Note that the marker allele names are the same in different markers, this does not mean that the alleles are the same, it just means that the notation used was to call alleles from *m1–mn*, where *n* is the total number of alleles. Note that microsatellite genotyping is much less automated than, e.g., SNP genotyping using arrays and labs tend to use ad-hoc formats for the data (e.g., allele names may be based on fragment size, an official nomenclature or some internal referencing system). In *prog* we already see some trouble. *Weight* which should be a numeric trait is being treated as a factor, that's because the notation used for missing is "-". We'll have to fix this. With the markers we also see that there's missing data, again using "-". There also seem to be more alleles than in the sire group—they must be coming from the dams. Let's see what the missing weight looks like, for this we can use the function *which*

```
> prog[which(prog$weight=="-"),]
        id  sire sex weight m11 m12 m21 m22 m31 m32 m41 m42
90 id90 sire3   M      -   M4  M1  M2  M5  M2  M5  M4  M2
    m51 m52
90   M1  M1
```

which returns an index of the position of the items true for the logical test. The logical operators in R are

- *equality* ==
- *not equal !=*
- *greater than* > *or greater than or equal* >=
- *smaller than* < *or smaller than or equal* <=

There's a single missing record for weight. Might as well just remove the entire record straight away, there's not much to associate a missing record to!

```
> prog=prog[-which(prog$weight=="-"),]
> prog$weight=as.numeric(as.character(prog$weight))
> summary(prog$weight)

   Min. 1st Qu.  Median    Mean 3rd Qu.    Max.
  282.8   307.0   331.0   329.9   351.8   480.5

> summary(sires$weight)

   Min. 1st Qu.  Median    Mean 3rd Qu.    Max.
  333.5   351.8   359.4   356.9   363.9   384.0
```

That looks better! We used the minus sign and *which* to remove the missing data, then converted our *weights* to a numeric value and then summarized them.

XY plot of weight by animal

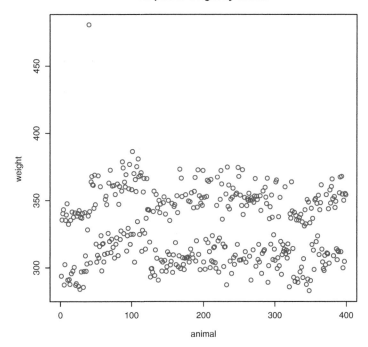

Fig. 2.2 Plot of weights—notice the outlier point and the clear separation of the data into two groups

If we compare the means and spread with the *sires* data we observe that the maximum value is much larger than that of the sires. It's probably a good idea to have a look at it.

```
> plot(prog$weight,main="XY plot of weight by animal",
+      xlab="animal",ylab="weight",col="blue")
```

Figure 2.2 shows that there is one measurement that is an outlier and the data seem to be separating very clearly into two groups—we have to keep this in mind.

We could handle the outlier in two ways—go back to the researchers and ask them about it: does it really make sense, can they go back to the original data and check it out. Or just remove it from our dataset, let's do that. From Fig. 2.2 we know that the outlier is above 400, we can use *which* again

```
> prog=prog[-which(prog$weight>400),]
> summary(prog$weight)
```

```
   Min. 1st Qu.  Median    Mean 3rd Qu.    Max.
  282.8   307.0   330.1   329.5   351.6   386.3
```

Now let's have a look at the marker data and see which genotypes are missing

```
> index=grep("m",names(prog))
> missing=numeric()
> for (i in 1:length(index))
+       missing=c(missing,which(prog[,index[i]]=="-"))
> print(missing)

 [1]  69 69 69 69 69 69 69 69 69 69

> missingU=unique(missing)
> print(missingU)

[1]  69
```

There's quite a lot going on here. First we used the function *grep* that searches for pattern matches—it uses regular expressions and it's a very powerful tool. What we did was search for the letter *m* in the header names of *prog*, this returns the indexes of the markers. Of course this is overkill here and using a single letter for the searches is rather dangerous, only do this if you really know the header structure. The point was just to call attention to the function; try to spend some time learning how to use *grep*, it is really handy to find patterns in data. Next we created a new variable called *missing*, and defined it as a numeric variable. We had to declare the variable explicitly because in the next line we use a *loop* and R will not assign values to a variable in a loop unless it has been previously declared. The syntax for loops in R is *for(i in x:y)*, *i* is the counter, *x* is the starting value and *y* is the final value. The value of *i* will change in unit increments unless you explicitly modify it. Inside the loop instead of a hard coded value we used *length* to find out the actual number of marker columns—just makes the code more flexible. We used a loop to go over all marker columns and added to *missing* the indices of the marker alleles with "-".

When we print the indices in *missing* and eyeball the results, we see that there is a single index repeated across all markers. We can also use the function *unique* to get only the unique non-repeated values in *missing*. Now let's look at the missing genotype

```
> print(prog[missingU,])
      id   sire sex weight m11 m12 m21 m22 m31 m32 m41 m42
70 id70 sire2   M 372.45   -   -   -   -   -   -   -   -
    m51 m52
70    -   -
```

That's great! The same animal failed genotyping across all markers/alleles and all other records are complete—probably just a bad DNA sample. We can again exclude this record using *which* and the index of the missing data

```
> prog=prog[-missingU,]
> summary(prog$m11)
```

```
-    M1   M2   M3   M4   M5
0   97  139  101   59    1
```

Note that "-" is still there as a level with no records. Probably not necessary but we can remove it by re-leveling the factors (will only keep levels with records)

```
> for (i in 1:length(index))
+       prog[,index[i]]=factor(prog[,index[i]])
> summary(prog$m11)
```

```
M1   M2   M3   M4   M5
97  139  101   59    1
```

And we have cleaned up our data. This is rather convoluted and there are better ways of doing it (e.g., declare "-" as the symbol for missing data when reading in the files, use *na.strings*="-" as an argument in *read.table*). We are almost ready for the analysis, but first let's try to find out why our *weight* data is separating so distinctly into two groups. Our data only has two potential fixed effects—*sire* and *sex*. Let's plot *weights* grouped by these two effects

```
> boxplot(prog$weight~prog$sire,
+       col=1:length(levels(prog$sire)),
+       main="Boxplot of weights by sire")
> boxplot(prog$weight~prog$sex,
+       col=1:length(levels(prog$sex)),
+       main="Boxplot of weights by sex")
```

Figure 2.3 does not show too much difference due to *sires*. Now, the effect of *sex* in Fig. 2.4 is really large, that's were our bimodality (Fig. 2.5) is coming from

```
> plot(density(prog$weight),col="blue",
+       main="Density plot of weights")
```

We can show that it really is *sex* that is splitting our data by repeating the previous XY plot using *sex* to color the points (Fig. 2.6). Note that we also used the argument *pch* to change the symbol used to plot each data point and used *legend* so we know what is what. The arguments for legend are position (*topleft, bottomright*, etc.), the items [here we used directly the levels in *prog$sex* but could have used, e.g., *c("female","male")* instead], the colors (first two in the active palette), and likewise *pch* for the symbols.

```
> plot(prog$weight,col=prog$sex, pch=as.numeric
  (prog$sex),
+       main="XY plot of weight by animal",
+       xlab="animal",ylab="weight")
>   legend("topleft",levels(prog$sex),col=1:2,pch=1:2)
```

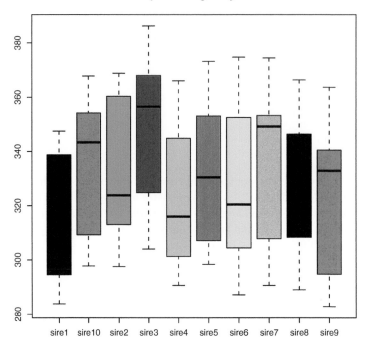

Fig. 2.3 Boxplot of weights by sire

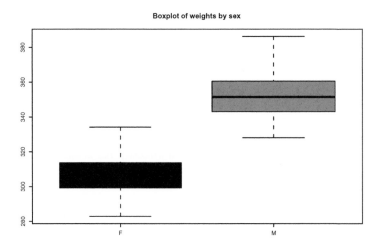

Fig. 2.4 Boxplot of weights by sex

Density plot of weights

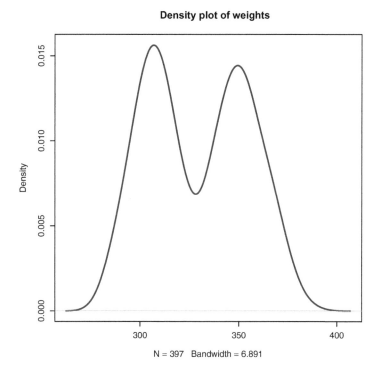

N = 397   Bandwidth = 6.891

Fig. 2.5 Density plot of weights

In our dataset we have genotypes for sires and offspring. One more test we can do is to see if there are any genotyping errors that we can pick up by comparing the genotypes of the sires with the genotypes of the offspring. We would expect that the most common alleles in the offspring would be those that show up in the sire. In the code below we use *grep* again to get the indices for the markers in both *sires* and *prog*, then we "*bend*" these indices into a matrix because we have one column for one allele and then the next column for the second allele. Next we use *library* to load a package called *made4* (note: to load packages use *library(packagename)* but it needs to be already installed in the library folder). Here we are using a function (*comparelists*) in the package to compare two vectors. A new variable *compatible* is created to hold the results—the first argument is to say that we want to fill the matrix with *NA*, then the dimensions of the matrix (number of sires by number of markers). The next step is a loop (actually two loops—the first one goes over the markers, the second one over the sires). For each sire and each marker, we get the alleles of the sire (variable *sirealleles*), we summarize the allele counts of the offspring of the sire, sort these counts in decreasing order and assign the top two to *topalleles*. Then we compare *sirealleles* with *topalleles* using *comparelists* and assign the length of the differences (number of alleles that do not match between the two) to the correct index in *compatible*. In this manner we have gone over all sires and all markers and have a matrix in which any number that is not *0* could be an indication of problems.

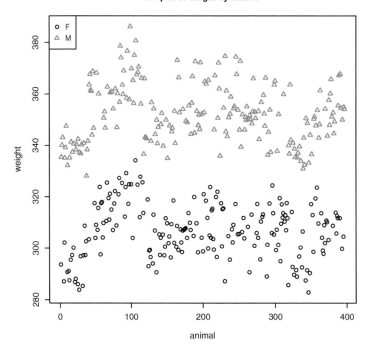

Fig. 2.6 Plot of weights color coded by sex

```
> indexms=grep("m",names(sires))
> indexms=matrix(indexms,length(indexms)/2,2,byrow=T)
> print(indexms)

      [,1] [,2]
[1,]     3    4
[2,]     5    6
[3,]     7    8
[4,]     9   10
[5,]    11   12

> indexm=grep("m",names(prog))
> indexm=matrix(indexm,length(indexm)/2,2,byrow=T)
> print(indexm)

      [,1] [,2]
[1,]     5    6
[2,]     7    8
[3,]     9   10
[4,]    11   12
[5,]    13   14
```

```
> library(made4)
> compatible=matrix(NA,length(sires$id),
+ length(indexms[,1]))
> for (j in 1:length(indexms[,1]))
+ {
+   for (i in 1:length(sires$id))
+   {
+     indexs=which(prog$sire==sires$id[i])
+
+     sirealleles=sires[i,indexms[j,]]
+     sirealleles=c(as.character(sirealleles[,1]),
+         as.character(sirealleles[,2]))
+
+     hold=prog[indexs,indexm[j,]]
+     hold=factor(c(as.character(hold[,1]),
+         as.character(hold[,2])))
+     hold=sort(summary(hold),decreasing=T)
+     topalleles=names(hold)[1:2]
+
+     compatible[i,j]=
+     length(comparelists(sirealleles,topalleles)$Set
+      .Dif)
+
+     if(i==1 & j==1)
+     {
+       cat("allele counts in offspring\n")
+       print(hold)
+       cat("most common alleles in offspring\n")
+       print(topalleles)
+       cat("sire alleles\n")
+       print(sirealleles)
+     }
+   }
+ }
allele counts in offspring
M2 M1 M3 M5 M4 M6
27 26  9  8  6  2
most common alleles in offspring
[1] "M2" "M1"
sire alleles
[1] "M2" "M1"

> print(compatible)
```

	[,1]	[,2]	[,3]	[,4]	[,5]
[1,]	0	0	0	0	0
[2,]	0	0	0	0	0
[3,]	0	0	0	0	0
[4,]	0	0	0	0	0
[5,]	0	0	0	0	0
[6,]	0	0	0	0	0
[7,]	0	0	0	0	0
[8,]	0	0	0	0	0
[9,]	0	0	0	0	0
[10,]	0	0	0	0	0

Our matrix *compatible* is filled with zeros. Which means we could not pick up any inconsistency between sire and offspring genotypes. Our conclusion here is that the sires were probably genotyped correctly. Unfortunately, real life is much more complicated, the results will not look so neat if, e.g., there are few offspring per sire, but I hope this helps to get you started.

Note the use of the *if* statement. It was used so that the contents of the variables were printed only once—one sire and one marker (could use this for checking purposes). The symbol & is the logical *and* (e.g., *IF this is TRUE AND that is also TRUE then do...*). By the way the symbol for logical *or* is |. The function *cat* is similar to print, but a bit prettier!

There are many important aspects in this code. We used a function from a library (in practice could also have simply used the equivalent function *setdiff* from the base installation); built and populated matrices, illustrated the usage of double loops (rather inefficient in R but handy with not too large datasets), used *if* to test a condition and run some other code, used nested functions to collate results (e.g., *sort(summary(hold))*).

Now, returning to the problem, let's flip the question. Given that the sire genotypes are correct, can we pick up any offspring that might be incorrectly genotyped? We do basically the same but this time we compare the alleles of the sire with the alleles in each offspring and check for Mendelian inconsistencies— at least one of the alleles should be common between the two. There might have been a mutation but it's much more likely that it's a genotyping error. This works better with markers that have many alleles, if the dam had the same allele(s) as the sire we would not be able to pick up errors as the offspring would appear to be consistent with the sire but in reality received the allele from the dam's side.

The code is similar to what we did before. Variable *compatible* (note that we are overwriting the previous matrix) now has one row for each offspring and again one column for each marker. We also use an additional loop to go over all offspring for a given sire. Note at the end that we convert *compatible* to a data.frame and then to factors (makes the summary more meaningful). We need to first convert the matrix to a data.frame to be able to change the type of data (from numeric to factor, a matrix will not let you change the data type on a column by column basis—a matrix can only hold a single data type).

```
> compatible=matrix(NA,length(prog$id),
+       length(indexms[,1]))
> for (j in 1:length(indexms[,1]))
+ {
+    for (i in 1:length(sires$id))
+    {
+        indexs=which(prog$sire==sires$id[i])
+
+        sirealleles=sires[i,indexms[j,]]
+        sirealleles=c(as.character(sirealleles[,1]),
+            as.character(sirealleles[,2]))
+
+        for (k in 1:length(indexs))
+        {
+          hold=prog[indexs[k],indexm[j,]]
+          topalleles=c(as.character(hold[,1]),
+            as.character(hold[,2]))
+          compatible[indexs[k],j]=
+          length(comparelists(sirealleles,topalleles)
+        $intersect)
+        }
+    }
+ }
> compatible=data.frame(compatible)
> for (i in 1:length(compatible[1,]))
+     compatible[,i]=factor(as.character(compatible[,i]))
> cat("\nSummary of alleles in common between
+       sires and offspring\n")

Summary of alleles in common between
    sires and offspring

> summary(compatible)

  X1          X2          X3          X4          X5
  0:   1      1:332       1:326       1:324       1:336
  1:333       2:  65      2:  71      2:  73      2:  61
  2:  63
```

It's looking good except for the first marker—one animal has no corresponding allele in its sire. For offspring that have a *2* (two alleles in common with the sire) we cannot determine which one came from the sire and which one came from the dam, since in our dataset the dams were not genotyped. There is a workaround using imputation. We could look at the frequencies of the alleles and try to impute the allele coming from the sire based on a probability function. Here our data is a bit too small for that and imputation only really becomes important when we have

missing data in our genotypes. But what we have to do is look into the record that
does not have any allele in common with the sire.

```
> index=which(compatible[,1]==0)
> print(prog[index,])

   id   sire sex weight m11 m12 m21 m22 m31 m32 m41 m42
1 id1 sire1   F 293.61  M5  M6  M2  M5  M3  M6  M2  M5
   m51 m52
1  M4  M4
```

Let's simply drop the record. This is of course overkill. You would not normally
drop the entire record (you'd just set this marker to *NA*), after all the other genotypes
seem to be ok. But since we did not have to pay for this data we can *afford* to be
over conservative and work with the assumption that if one genotype is not correct
we do not trust the others—guilt by association.

```
> prog=prog[-index,]
> dim(prog)

[1] 396   14
```

Four animal records were removed from our dataset—just one percent. But don't
expect such good data from real experiments.

The next thing we can do with the data is plot the genotypic and allelic
frequencies (Fig. 2.7). Let's work with data from the first marker across all families.
First we have to summarize the count data for alleles

```
> alleles=summary(factor(c(as.character(prog$m11),
+       as.character(prog$m12))))
> print(alleles)

 M1   M2   M3   M4   M5   M6
166  207  167  121   74   57

> alleles=alleles/sum(alleles)
> print(alleles)

        M1          M2          M3          M4          M5
0.20959596  0.26136364  0.21085859  0.15277778  0.09343434
        M6
0.07196970
```

That was easy. The function *sum* does what you'd expect. And now for
genotypes...

```
> hold=data.frame(m11=as.character(prog$m11),
+       m12=as.character(prog$m12))
> hold[,1]=as.character(hold[,1])
```

Barplot of allelic frequencies

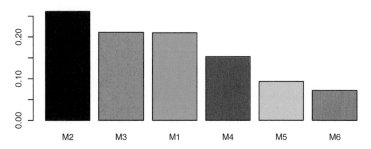

Barplot of genotypic frequencies

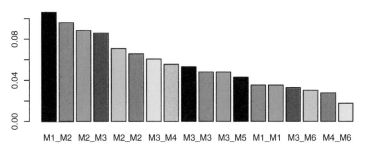

Fig. 2.7 Plot of allelic and genotypic frequencies for marker 1

```
> hold[,2]=as.character(hold[,2])
> sorted=character()
> for (i in 1:length(hold[,1]))
+     sorted=rbind(sorted,sort(as.character(hold[i,])))
> genotypes=paste(as.character(sorted[,1]),
+     as.character(sorted[,2]),sep="_")
> genotypes=summary(factor(genotypes))
> print(genotypes)

M1_M1 M1_M2 M1_M3 M1_M4 M1_M5 M1_M6 M2_M2 M2_M3 M2_M4
   14    42    34    22    26    14    28    35    38
M2_M5 M2_M6 M3_M3 M3_M4 M3_M5 M3_M6 M4_M4 M4_M5 M4_M6
   17    19    21    24    19    13     7    12    11

> genotypes=genotypes/sum(genotypes)
> print(genotypes)

     M1_M1        M1_M2        M1_M3        M1_M4        M1_M5
0.03535354 0.10606061 0.08585859 0.05555556 0.06565657
```

```
         M1_M6           M2_M2           M2_M3           M2_M4           M2_M5
0.03535354  0.07070707  0.08838384  0.09595960  0.04292929
         M2_M6           M3_M3           M3_M4           M3_M5           M3_M6
0.04797980  0.05303030  0.06060606  0.04797980  0.03282828
         M4_M4           M4_M5           M4_M6
0.01767677  0.03030303  0.02777778
```

That was much less simple. The problem here is that we have to merge the two
alleles together to form the genotype. And not only that, we have to make sure that
AB is not considered different from *BA*. First we created a temporary data.frame to
hold the two alleles, then a new variable to hold the sorted order of the alleles was
used—*sorted*. Next we sorted each pair of alleles using a loop and added the sort
order to *sorted*. Finally, we merged the sorted alleles into a single genotype using
the function *paste* which concatenates strings. The rest is the same as we did with
the alleles.

Now for plotting. The functions *split.screen* and *screen* are a handy way of
placing more than one image in a single plot area (instead of having to make two
separate plots, see Fig. 2.7).

```
> split.screen(c(2,1))

[1] 1 2

> screen(1)
> barplot(sort(alleles,decreasing=T),col=1:11,
+       main="Barplot of allelic frequencies")
> screen(2)
> barplot(sort(genotypes,decreasing=T),col=1:11,
+       main="Barplot of genotypic frequencies")
> close.screen(all = TRUE)
> dev.print(file="images/animgeno.pdf",
+       device=pdf,width=8,height=8)
```

We sorted the frequencies in decreasing order before plotting. We can see that the
most frequent alleles are the ones coming from the sires. Recall the use of *dev.print*
for saving plots.

For the analysis we will use a genotype model. Let's make a new data.frame with
the same information but with a single column for the genotypes for each marker
instead of two columns with alleles. We can do exactly the same as before and just
add a loop and a new variable *allgeno* to store the genotypes. Then we create a new
data.frame with the progeny data and the genotypes called *markers*.

```
> allgeno=NULL
> for(i in 1:length(indexm[,1]))
+ {
+    hold=data.frame(prog[indexm[i,]])
+    hold[,1]=as.character(hold[,1])
```

```
+     hold[,2]=as.character(hold[,2])
+     sorted=character()
+     for (i in 1:length(hold[,1]))
+        sorted=rbind(sorted,sort(as.character(hold[i,])))
+     genotypes=paste(as.character(sorted[,1]),
+        as.character(sorted[,2]),sep="_")
+     allgeno=cbind(allgeno,genotypes)
+ }
> colnames(allgeno)=c("M1","M2","M3","M4","M5")
> markers=data.frame(prog[,1:4],allgeno)
> head(markers)
```

```
    id  sire sex weight     M1     M2     M3     M4     M5
2  id2 sire1   M 335.43 M1_M4 M1_M3 M4_M5 M3_M4 M2_M2
3  id3 sire1   M 340.09 M2_M3 M2_M6 M1_M3 M3_M4 M3_M4
4  id4 sire1   M 343.08 M2_M3 M1_M2 M4_M6 M2_M3 M4_M5
5  id5 sire1   F 287.08 M1_M3 M2_M4 M4_M6 M4_M5 M2_M3
6  id6 sire1   F 302.17 M2_M2 M2_M5 M3_M5 M2_M4 M1_M4
7  id7 sire1   M 335.11 M1_M2 M1_M2 M3_M3 M2_M4 M2_M5
```

Now that we have cleaned the data, let's save it so that if we decide to rerun the analysis later on we don't need to go through all the cleaning steps again.

```
> write.table(markers,"chapter2/cleandata.txt",
+ quote=F,sep="\t",row.names=F)
```

At long last we are ready to analyze the data. This is quite common with genomic data (and almost all statistical analyses for that matter), more than half of the work involves cleaning up and preprocessing.

We want to test the markers with a continuous trait (*weight*) for association. A good way to go about it is with a linear model for analysis of variance (anova). Here is where R comes into its own. We can use the function *lm* to fit models, it can handle regressions, analysis of variance and analysis of covariance. The syntax is *lm(formula=mymodel, data=mydata)*, there are other arguments such as how to handle missing data, what weights should be used in the fitting process, you can define a specific subset of the data to be used and also a list of contrasts to be tested (see the help page for details). The argument *data* is just the name of the data.frame that will be used in the analysis. Instead of having to use the full name (*data.frame$headername*) as we have been using so far, you can then use only the header names in the formula and these will be picked up from the data.frame specified in *data*.

A side note here—if you don't want each time to write the name of the data.frame and the name of the header you can use *attach(data.frame)*, this will make all headers in the data.frame directly available in R.

```
> attach(sires)
> print(id)
```

```
[1] sire1   sire2   sire3   sire4   sire5   sire6   sire7
[8] sire8   sire9   sire10
10 Levels: sire1 sire10 sire2 sire3 sire4 ... sire9

> detach(sires)
```

Just be careful, for example if you attach both *sires* and *prog* you will run into trouble because of the names in common that are in both data.frames. It's probably best not to attach a data.frame but if you do, remember to *detach* it when you are finished—this removes the *direct access* and its consequences from R's workspace. To exemplify the problem

```
> attach(sires)
> attach(prog)

The following object(s) are masked from 'sires':

    id, m11, m12, m21, m22, m31, m32, m41, m42,
    m51, m52, weight

> detach(sires)
> detach(prog)
```

Now, the important part is our model formula. The syntax is somewhat similar to how we would normally write a model $y \sim x_1 + x_2 + x_n$. As we mentioned before, the tilde symbol ˜ means "modeled by". On the left-hand side we have our response variable and on the right-hand side our explanatory variables. For our data we could test something along the lines of *weight ˜ M1*. This models weight in relation to the genotypes of marker 1. The full R syntax is

  lm(weight ˜ M1, data=markers)

Let's just run the model in R and see what happens

```
> lm(weight~M1,data=markers)

Call:
lm(formula = weight ~ M1, data = markers)

Coefficients:
(Intercept)        M1M1_M2        M1M1_M3        M1M1_M4
  3.253e+02      1.771e+00      2.256e+00      8.919e+00
    M1M1_M5        M1M1_M6        M1M2_M2        M1M2_M3
 -1.372e+00      7.143e-04     -1.049e+00      9.872e+00
    M1M2_M4        M1M2_M5        M1M2_M6        M1M3_M3
 -3.144e-01     -2.640e+00      1.145e+01      4.787e+00
    M1M3_M4        M1M3_M5        M1M3_M6        M1M4_M4
  7.156e+00      1.295e+01      8.932e+00      1.955e+01
    M1M4_M5        M1M4_M6
  1.065e+01      6.623e-02
```

R prints out the model formula (*call*) and the coefficients. That's nice but not quite enough. Let's assign the results to a variable and see what that looks like

```
> results=lm(weight~M1,data=markers)
> class(results)

[1] "lm"
```

Our variable *results* is of class *lm*, which is a class that holds results from a linear model and has quite a few methods that can be applied to it. The first and most important method is our already familiar *summary*.

```
> summary(results)

Call:
lm(formula = weight ~ M1, data = markers)

Residuals:
    Min      1Q  Median      3Q     Max
-46.185 -22.118   1.607  21.368  52.109

Coefficients:
              Estimate Std. Error t value Pr(>|t|)
(Intercept)  3.253e+02  6.763e+00  48.095   <2e-16 ***
M1M1_M2      1.771e+00  7.809e+00   0.227   0.8207
M1M1_M3      2.256e+00  8.035e+00   0.281   0.7790
M1M1_M4      8.919e+00  8.651e+00   1.031   0.3032
M1M1_M5     -1.372e+00  8.388e+00  -0.164   0.8702
M1M1_M6      7.143e-04  9.564e+00   0.000   0.9999
M1M2_M2     -1.049e+00  8.283e+00  -0.127   0.8993
M1M2_M3      9.872e+00  8.002e+00   1.234   0.2181
M1M2_M4     -3.144e-01  7.911e+00  -0.040   0.9683
M1M2_M5     -2.640e+00  9.132e+00  -0.289   0.7727
M1M2_M6      1.145e+01  8.913e+00   1.284   0.1998
M1M3_M3      4.787e+00  8.731e+00   0.548   0.5838
M1M3_M4      7.156e+00  8.510e+00   0.841   0.4009
M1M3_M5      1.295e+01  8.913e+00   1.453   0.1472
M1M3_M6      8.932e+00  9.746e+00   0.916   0.3600
M1M4_M4      1.955e+01  1.171e+01   1.669   0.0959 .
M1M4_M5      1.065e+01  9.954e+00   1.069   0.2856
M1M4_M6      6.623e-02  1.020e+01   0.006   0.9948
---
Signif. codes:  0 '***' 0.001 '**' 0.01 '*' 0.05 '.'

Residual standard error: 25.3 on 378 degrees of freedom
Multiple R-squared: 0.0432, Adjusted R-squared:
```

```
0.0001707 F-statistic: 1.004 on 17 and 378 DF,
  p-value: 0.4529
```

Now we get the model formula, the range of the residuals of our fitted data and the coefficients with the estimates for each class, the standard errors of the estimates, the *t*-values and the *p*-values. And also the usual *significance stars*. We also get some additional information regarding residual standard errors and number of degrees of freedom, goodness of fit of the model with the *R*-squared and adjusted *R*-squared, and so on. We will only focus on the interpretation of the output, readers interested in a step by step walkthrough of the calculations might enjoy the introductory book by Crawley [21] or, for a more in-depth theoretical exposition, Faraway's [31] book is a good reference.

It's always a good idea to have a quick look at the (adjusted) *R* squared or the *p*-value of the *F*-statistic. It's not necessarily decisive in model selection but it does provide a broad indication if the model is actually explaining any part of the observations or not. In our example we can see that it's a rather poor fit—the adjusted *R*-squared is almost zero. It's also a good idea to check if the model is behaving sensibly (if our assumptions are not too strong—e.g., data is normal, residuals are well behaved, variance is not changing based on our measures). Use *plot(lmobject)* to get four "*quality control*" plots of your model (there are actually six plots, but four are shown by default). The first two are the most interesting, an *xy* plot of fitted by residual values and a QQ plot of theoretical normal quantiles by residual quantiles. Just using plot on an *lmobject* is the most convenient way for plotting, but you can also make the plots yourself. To extract the fitted values from an *lmobject* use the function *predict* and for the residuals the function *residuals* (that's another two more methods for an *lmobject*), then just plot one against the other

```
> plot(predict(results),residuals(results),
+      main="XY plot of residuals X fitted values",
+      xlab="fitted values (weight)",
+      ylab="residuals",col="blue")
```

We do not observe too much of a pattern in the plot (Fig. 2.8) so the variance is constant (or at least constant enough). For a normal QQ plot it's as simple as

```
> split.screen(c(2,1))
> screen(1)
> qqnorm(predict(results),col="blue")
> screen(2)
> qqnorm(residuals(results),col="blue",main="")
> close.screen(all = TRUE)
```

The top plot in Fig. 2.9 shows the fitted values against the normal theoretical quantiles and the lower plot shows the same but for the residuals (which would be the second plot given by *plot(results)*). Note that in terms of normality we have nothing to write home about. I'm sure we can blame the effect of *sex* for it!

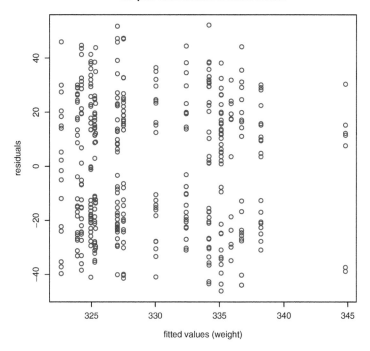

Fig. 2.8 XY plot of fitted values versus residuals. Note that the variance seems stable, i.e., independent of the values for *weight*

Another two functions that come in handy to check the data are *fligner.test* for homogeneity of variances and *shapiro.test* for normality.

```
> fligner.test(weight~M1,data=markers)

        Fligner-Killeen test of homogeneity of variances

data:  weight by M1
Fligner-Killeen:med chi-squared = 7.6979, df = 17,
p-value = 0.9726

> shapiro.test(prog$weight)

        Shapiro-Wilk normality test

data:  prog$weight
W = 0.9483, p-value = 1.548e-10
```

Once we are satisfied that nothing too amiss is going on with our model, we can interpret the results. In the summary table we have *Estimate, Std. Error, t value,* and *Pr(>|t|)*. It's quite straightforward. The estimate is the estimated *mean* value for

Normal Q-Q Plot

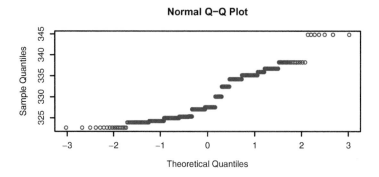

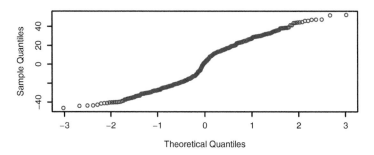

Fig. 2.9 QQ plot (observed versus theoretical normal quantiles). *Top plot*: fitted values. *Lower plot*: residual values

each factor. Note that what is called *Intercept* is the actual estimate for one of the levels and all other values are shown as a deviation from this top value (R by default uses a treatment–contrast parameterization). The baseline is selected in alphabetical order (in our example M1_M1), so if you want a different baseline you'll need to reorder the levels. The standard errors are, as you would expect, the uncertainty around the estimates and the last column shows the *p*-values. Note that the *p*-values are always in relation to the baseline factor.

Another useful function is *anova* which is essentially just a different output for an *lmobject* as a standard anova table (a good way to quantify the effect of explanatory variables).

```
> anova(results)

Analysis of Variance Table

Response: weight
           Df Sum Sq Mean Sq F value Pr(>F)
M1         17  10928  642.83   1.004 0.4529
Residuals 378 242028  640.29
```

The marker is not significant (does not explain much of the variability in weight). We know this model has little explanatory power, but we also know that *sex* seems to have a pretty big effect on our data. Let's include *sex* in the model

```
> summary(lm(weight~sex+M1,data=markers))

Call:
lm(formula = weight ~ sex + M1, data = markers)

Residuals:
      Min       1Q    Median       3Q       Max
 -23.1174   -7.2400   -0.7689   7.1190   31.3916

Coefficients:
              Estimate Std. Error t value Pr(>|t|)
(Intercept) 2.992e+02  2.900e+00 103.168  < 2e-16 ***
sexM        4.558e+01  1.081e+00  42.143  < 2e-16 ***
M1M1_M2     6.111e+00  3.274e+00   1.867 0.062693 .
M1M1_M3     5.512e+00  3.368e+00   1.637 0.102537
M1M1_M4     1.010e+01  3.625e+00   2.787 0.005590 **
M1M1_M5     1.884e+00  3.516e+00   0.536 0.592370
M1M1_M6     7.143e-04  4.007e+00   0.000 0.999858
M1M2_M2     7.090e+00  3.476e+00   2.040 0.042079 *
M1M2_M3     5.965e+00  3.354e+00   1.779 0.076118 .
M1M2_M4     7.739e+00  3.320e+00   2.331 0.020292 *
M1M2_M5     4.637e+00  3.830e+00   1.211 0.226794
M1M2_M6     1.350e+01  3.735e+00   3.615 0.000341 ***
M1M3_M3     9.128e+00  3.660e+00   2.494 0.013051 *
M1M3_M4     1.421e+01  3.569e+00   3.981 8.24e-05 ***
M1M3_M5     1.260e+01  3.734e+00   3.375 0.000815 ***
M1M3_M6     1.043e+01  4.084e+00   2.555 0.011010 *
M1M4_M4     1.304e+01  4.910e+00   2.655 0.008258 **
M1M4_M5     1.010e+01  4.171e+00   2.422 0.015898 *
M1M4_M6     9.537e+00  4.278e+00   2.229 0.026375 *
---
Signif. codes:  0 '***' 0.001 '**' 0.01 '*' 0.05 '.'

Residual standard error: 10.6 on 377 degrees of freedom
Multiple R-squared: 0.8325, Adjusted R-squared: 0.8245
F-statistic: 104.1 on 18 and 377 DF, p-value: < 2.2e-16
```

That is a big effect! It is highly significant and we expect males to be around *46 kg* heavier than females. That's a lot more than any of the other effects. And the adjusted *R*-squared is also looking much better at over 0.82 (largely due to *sex*). And now a lot of genotypes seem to have an effect. So, that's that for marker 1? Not quite. We have not considered the effect of *sire* in our model

```
> summary(lm(weight~sex+sire+M1,data=markers))
```

Call:
lm(formula = weight ~ sex + sire + M1, data = markers)

Residuals:
     Min   1Q   Median   3Q   Max
-20.5650   -5.7110   -0.0889   5.2261   21.5778

Coefficients:
              Estimate Std. Error t value Pr(>|t|)
(Intercept) 293.8628   2.6104 112.574 < 2e-16 ***
sexM   45.5573   0.8566   53.186 < 2e-16 ***
siresire10   14.5459   1.9541   7.444 7.00e-13 ***
siresire2   18.2750   2.0120   9.083 < 2e-16 ***
siresire3   27.9691   2.0110   13.908 < 2e-16 ***
siresire4   8.5190   1.9165   4.445 1.16e-05 ***
siresire5   14.2383   1.9915   7.150 4.72e-12 ***
siresire6   15.8241   2.0043   7.895 3.37e-14 ***
siresire7   14.2136   2.0673   6.875 2.66e-11 ***
siresire8   14.1905   2.1265   6.673 9.22e-11 ***
siresire9   3.6338   1.9234   1.889   0.0596 .
M1M1_M2   1.0713   2.6239   0.408   0.6833
M1M1_M3   -0.7715   2.7549   -0.280   0.7796
M1M1_M4   0.9751   3.0089   0.324   0.7461
M1M1_M5   -1.6311   2.8036   -0.582   0.5611
M1M1_M6   -1.6844   3.1542   -0.534   0.5937
M1M2_M2   -0.4491   2.8385   -0.158   0.8744
M1M2_M3   -2.0030   2.7399   -0.731   0.4652
M1M2_M4   -1.4636   2.7087   -0.540   0.5893
M1M2_M5   -1.9519   3.0634   -0.637   0.5244
M1M2_M6   2.4693   3.0634   0.806   0.4207
M1M3_M3   -0.8174   3.0775   -0.266   0.7907
M1M3_M4   2.1514   2.9849   0.721   0.4715
M1M3_M5   2.2666   3.1662   0.716   0.4745
M1M3_M6   0.4037   3.3847   0.119   0.9051
M1M4_M4   -3.6682   4.0730   -0.901   0.3684
M1M4_M5   1.2526   3.6037   0.348   0.7283
M1M4_M6   -3.0625   3.6157   -0.847   0.3975

Signif. codes: 0 '***' 0.001 '**' 0.01 '*' 0.05 '.'

Residual standard error: 8.303 on 368 degrees of freedom
Multiple R-squared: 0.8997, Adjusted R-squared: 0.8923
F-statistic: 122.3 on 27 and 368 DF, p-value: < 2.2e-16
```
```

Sire also seems to have quite a large influence on our data and note that the variation that before was attributed to the marker is now being captured by the sires. Be careful when interpreting these *p*-values, remember that they reflect differences from the base level (here *sire 1*). We could have a scenario with all sires significantly different from *sire 1* but not different between themselves. They would all seem to be significant but the actual variation explained by sire might not be. For example, *sires 5, 7 and 8* are all quite similar to each other (the estimated difference between *sires 7 and 8* is only 22.1 g—14.2136 − 14.1905 = 0.0221). Note that *sire 3* has the highest *p*-value (0.0596) and is also the one with the smallest difference (3.6338 kg) from *sire 1*. Coincidentally *sire 1* has the lightest offspring (on average), hence all coefficients are positive. So, these *p*-values do not tell us how much of the variation is explained by the sires and we still do not know if the term is significant enough to be worthwhile keeping in the model. To formally test the difference in models (essentially proportion of variation explained by a term) do an anova on them.

```
> model1=lm(weight~sex+sire+M1,data=markers)
> model2=lm(weight~sex+M1,data=markers)
> model3=lm(weight~M1,data=markers)
> anova(model1,model2)

Analysis of Variance Table

Model 1: weight ~ sex + sire + M1
Model 2: weight ~ sex + M1
  Res.Df    RSS Df Sum of Sq       F    Pr(>F)
1    368  25370
2    377  42379 -9    -17009  27.413 < 2.2e-16 ***
---
Signif. codes:  0 '***' 0.001 '**' 0.01 '*' 0.05 '.'

> anova(model2,model3)

Analysis of Variance Table

Model 1: weight ~ sex + M1
Model 2: weight ~ M1
  Res.Df     RSS Df Sum of Sq       F    Pr(>F)
1    377   42379
2    378  242028 -1   -199649  1776.1 < 2.2e-16 ***
---
Signif. codes:  0 '***' 0.001 '**' 0.01 '*' 0.05 '.'
```

Sire has a highly significant effect in our model and the term should be kept. Out of completeness we also ran an anova test to see if we had to keep *sex* and of course, no questions there. In summary to test if a term should be kept or deleted from a model, build the models you want to try, store them as an *mobject* and then compare them using an anova. We can also test our model against a *null* model—a model that has no explanatory power at all.

```
> model4=lm(weight~1,data=markers)
> anova(model3,model4)

Analysis of Variance Table

Model 1: weight ~ M1
Model 2: weight ~ 1
  Res.Df    RSS Df Sum of Sq      F Pr(>F)
1    378 242028
2    395 252956 -17    -10928 1.004 0.4529
```

And this confirms it—our marker is not providing any information at all. But we already knew that from the *R*-squared and anova. Note that the *p*-value above (0.4529) and the *p*-values from the anova and from the *F*-statistic given by *summary* that we obtained with the model fitting only the marker are exactly the same—as they should be. In contrast look at *sex*

```
> anova(model2,model4)

Analysis of Variance Table

Model 1: weight ~ sex + M1
Model 2: weight ~ 1
  Res.Df    RSS Df Sum of Sq      F      Pr(>F)
1    377  42379
2    395 252956 -18   -210577 104.07 < 2.2e-16 ***
---
Signif. codes:  0 '***' 0.001 '**' 0.01 '*' 0.05 '.'
```

So far we have not mentioned interactions between terms. If we wanted a fully saturated model with all terms and interactions (and completely inestimable from this data) it would look like

*weight ˜ sex*sire*M1*

The symbol * is used to model the main effect and the interactions. To estimate only the interactions use *sex:sire*. For quadratic terms use ^2, consult the R help files for details. This analysis was for the first marker across all families (but we did include *sire* in the model). We could also look at the effects on a per *sire*, single family model. For the first *sire* it would be

```
> summary(lm(weight~sex+M1,
+      data=markers[which(prog$sire=="sire1"),]))

Call:
lm(formula = weight ~ sex + M1, data = markers[which
  (prog$sire =="sire1"), ])

Residuals:
```

```
     Min       1Q    Median       3Q       Max
 -9.1959  -3.9096    0.0613   3.8731  12.6641
```

Coefficients:

```
             Estimate Std. Error t value Pr(>|t|)
(Intercept) 292.0005     3.9627  73.687   <2e-16 ***
sexM         44.0293     2.4834  17.730   <2e-16 ***
M1M1_M2       4.0354     4.1938   0.962    0.344
M1M1_M3      -0.7563     4.6992  -0.161    0.873
M1M1_M4      -0.5998     7.2481  -0.083    0.935
M1M1_M5       2.6276     4.7673   0.551    0.586
M1M1_M6       4.2302     7.2481   0.584    0.564
M1M2_M2       1.7728     5.3540   0.331    0.743
M1M2_M3       5.1777     4.8342   1.071    0.294
M1M2_M4       1.0178     4.6020   0.221    0.827
M1M2_M5      -0.1500     5.0916  -0.029    0.977
---
Signif. codes:  0 '***' 0.001 '**' 0.01 '*' 0.05 '.'
```

```
Residual standard error: 6.236 on 27 degrees of freedom
Multiple R-squared: 0.9484, Adjusted R-squared: 0.9292
F-statistic: 49.59 on 10 and 27 DF, p-value: 9.349e-15
```

The across families model is preferable—provided the marker and gene are fully linked, they will not recombine and the phases in the sire families are the same. If we run the same model with *sex* and *sire* for the other markers we get

```
> summary(lm(weight~sex+sire+M2,data=markers))
```

```
Call:
lm(formula = weight ~ sex + sire + M2, data = markers)
```

Residuals:

```
     Min       1Q    Median       3Q       Max
-22.7912  -5.6914   -0.5605   4.9794   23.5298
```

Coefficients:

```
             Estimate Std. Error t value Pr(>|t|)
(Intercept) 294.07371    2.56442 114.674  < 2e-16 ***
sexM         45.61291    0.86016  53.028  < 2e-16 ***
siresire10   15.96557    1.95229   8.178 4.74e-15 ***
siresire2    19.33987    1.94433   9.947  < 2e-16 ***
siresire3    27.56248    2.08319  13.231  < 2e-16 ***
siresire4     9.04794    2.01229   4.496 9.28e-06 ***
siresire5    14.91096    1.99842   7.461 6.23e-13 ***
siresire6    17.24295    1.96925   8.756  < 2e-16 ***
```

```
siresire7        15.51597       2.08658       7.436 7.36e-13 ***
siresire8        14.52662       2.04696       7.097 6.61e-12 ***
siresire9         3.64002       1.98009       1.838   0.0668 .
M2M1_M2          -1.35543       2.52170      -0.538   0.5912
M2M1_M3          -1.91512       2.40510      -0.796   0.4264
M2M1_M4          -4.73051       2.72847      -1.734   0.0838 .
M2M1_M5          -5.53041       2.96409      -1.866   0.0629 .
M2M1_M6           0.07772       2.74350       0.028   0.9774
M2M2_M2          -1.03401       2.71730      -0.381   0.7038
M2M2_M3          -1.36273       2.49493      -0.546   0.5853
M2M2_M4          -0.64443       2.51757      -0.256   0.7981
M2M2_M5           1.29490       2.85511       0.454   0.6504
M2M2_M6          -1.03280       2.65962      -0.388   0.6980
M2M3_M3          -1.90353       2.79155      -0.682   0.4957
M2M3_M4          -0.71610       2.68207      -0.267   0.7896
M2M3_M5          -4.12052       3.12912      -1.317   0.1887
M2M3_M6           2.27967       2.94271       0.775   0.4390
M2M4_M4           0.18093       2.94842       0.061   0.9511
M2M4_M5          -0.06594       3.82900      -0.017   0.9863
M2M4_M6           0.01959       3.13892       0.006   0.9950
---
Signif. codes:  0 '***' 0.001 '**' 0.01 '*' 0.05 '.'

Residual standard error: 8.285 on 368 degrees of freedom
Multiple R-squared: 0.9001, Adjusted R-squared: 0.8928
F-statistic: 122.9 on 27 and 368 DF, p-value: < 2.2e-16

> summary(lm(weight~sex+sire+M3,data=markers))

Call:
lm(formula = weight ~ sex + sire + M3, data = markers)

Residuals:
     Min        1Q    Median        3Q       Max
-23.3480   -5.8414   -0.3709    4.6576   23.8250

Coefficients:
            Estimate Std. Error t value Pr(>|t|)
(Intercept) 295.7821     3.4113  86.705  < 2e-16 ***
sexM         45.3620     0.8563  52.972  < 2e-16 ***
siresire10   14.5426     2.0227   7.190 3.65e-12 ***
siresire2    19.8409     2.0035   9.903  < 2e-16 ***
siresire3    27.9406     2.0000  13.970  < 2e-16 ***
siresire4     8.6615     1.9331   4.481 9.95e-06 ***
siresire5    15.1641     2.0205   7.505 4.67e-13 ***
siresire6    16.4759     1.9655   8.383 1.11e-15 ***
```

```
siresire7      15.0614      2.0775    7.250 2.48e-12 ***
siresire8      14.5809      1.9743    7.386 1.03e-12 ***
siresire9       3.2624      2.0031    1.629    0.104
M3M1_M2        -1.7039      3.4520   -0.494    0.622
M3M1_M3        -2.7986      3.3179   -0.843    0.400
M3M1_M4        -3.5924      3.5067   -1.024    0.306
M3M1_M5        -5.2669      3.7516   -1.404    0.161
M3M1_M6        -2.9969      3.9630   -0.756    0.450
M3M2_M2        -3.5294      3.7086   -0.952    0.342
M3M2_M3        -1.7399      3.3301   -0.522    0.602
M3M2_M4        -3.1061      3.3233   -0.935    0.351
M3M2_M5        -5.2805      3.5647   -1.481    0.139
M3M2_M6         2.7613      3.9215    0.704    0.482
M3M3_M3        -3.6043      3.6052   -1.000    0.318
M3M3_M4        -3.4710      3.3397   -1.039    0.299
M3M3_M5        -1.1773      3.6098   -0.326    0.745
M3M3_M6        -1.9418      3.5353   -0.549    0.583
M3M4_M4        -2.0071      3.8458   -0.522    0.602
M3M4_M5         0.4192      3.9026    0.107    0.915
M3M4_M6        -1.4906      3.8018   -0.392    0.695
---
Signif. codes:  0 '***' 0.001 '**' 0.01 '*' 0.05 '.'

Residual standard error: 8.303 on 368 degrees of freedom
Multiple R-squared: 0.8997, Adjusted R-squared: 0.8923
F-statistic: 122.3 on 27 and 368 DF, p-value: < 2.2e-16

> summary(lm(weight~sex+sire+M4,data=markers))

Call:
lm(formula = weight ~ sex + sire + M4, data = markers)

Residuals:
     Min       1Q    Median       3Q      Max
-22.7806  -5.8261   -0.0981   5.2882  23.2222

Coefficients:
             Estimate Std. Error t value Pr(>|t|)
(Intercept) 295.6093      3.1683  93.301  < 2e-16 ***
sexM         45.7384      0.8595  53.213  < 2e-16 ***
siresire10   15.7858      1.9570   8.066 1.03e-14 ***
siresire2    19.7087      1.9245  10.241  < 2e-16 ***
siresire3    28.1588      1.9848  14.187  < 2e-16 ***
siresire4     7.6661      2.0518   3.736 0.000216 ***
siresire5    14.7326      1.9683   7.485 5.33e-13 ***
siresire6    16.2506      2.0735   7.837 5.01e-14 ***
```

```
siresire7      14.5987      1.9936      7.323 1.54e-12 ***
siresire8      15.1862      1.9403      7.827 5.38e-14 ***
siresire9       3.1479      2.0130      1.564 0.118734
M4M1_M2        -4.0779      2.9327     -1.391 0.165216
M4M1_M3        -3.2466      3.1855     -1.019 0.308782
M4M1_M4        -2.2060      3.0873     -0.715 0.475355
M4M1_M5        -0.2960      3.4868     -0.085 0.932384
M4M1_M6         5.3292      3.8099      1.399 0.162723
M4M2_M2        -3.2301      3.3161     -0.974 0.330672
M4M2_M3        -4.0546      3.0258     -1.340 0.181061
M4M2_M4        -2.6226      3.0451     -0.861 0.389654
M4M2_M5        -2.0669      3.1708     -0.652 0.514888
M4M2_M6        -1.1982      3.2424     -0.370 0.711925
M4M3_M3        -1.9093      3.4852     -0.548 0.584141
M4M3_M4        -3.3156      3.1471     -1.054 0.292793
M4M3_M5        -0.9638      3.6675     -0.263 0.792853
M4M3_M6        -5.1306      3.7296     -1.376 0.169770
M4M4_M4         0.7503      3.4346      0.218 0.827189
M4M4_M5        -1.7691      3.6234     -0.488 0.625656
M4M4_M6        -6.6717      3.7661     -1.772 0.077298 .
---
Signif. codes:  0 '***' 0.001 '**' 0.01 '*' 0.05 '.'

Residual standard error: 8.226 on 368 degrees of freedom
Multiple R-squared: 0.9016, Adjusted R-squared: 0.8943
F-statistic: 124.8 on 27 and 368 DF, p-value: < 2.2e-16

> summary(lm(weight~sex+sire+M5,data=markers))

Call:
lm(formula = weight ~ sex + sire + M5, data = markers)

Residuals:
    Min      1Q  Median      3Q     Max
-14.192  -3.152  -0.124   3.451  13.538

Coefficients:
            Estimate Std. Error t value Pr(>|t|)
(Intercept)  305.860      1.813 168.731  < 2e-16 ***
sexM          46.151      0.519  88.921  < 2e-16 ***
siresire10    14.829      1.181  12.553  < 2e-16 ***
siresire2     10.591      1.299   8.151 5.72e-15 ***
siresire3     18.050      1.276  14.144  < 2e-16 ***
siresire4      6.169      1.200   5.139 4.48e-07 ***
siresire5     13.321      1.185  11.237  < 2e-16 ***
siresire6      9.559      1.283   7.449 6.77e-13 ***
```

```
siresire7       12.990      1.193   10.889  < 2e-16  ***
siresire8        5.610      1.334    4.205 3.29e-05  ***
siresire9        1.308      1.188    1.101   0.2717
M5M1_M2          1.505      1.804    0.834   0.4046
M5M1_M3          2.329      1.699    1.371   0.1713
M5M1_M4          3.405      1.794    1.898   0.0585  .
M5M1_M5          1.928      1.923    1.003   0.3166
M5M1_M6          2.232      1.852    1.206   0.2287
M5M2_M2        -16.271      2.123   -7.665 1.60e-13  ***
M5M2_M3        -14.598      1.728   -8.448 6.98e-16  ***
M5M2_M4        -14.061      1.795   -7.835 5.08e-14  ***
M5M2_M5        -14.299      1.937   -7.382 1.05e-12  ***
M5M2_M6        -17.000      2.091   -8.130 6.64e-15  ***
M5M3_M3        -14.144      1.869   -7.566 3.12e-13  ***
M5M3_M4        -13.574      1.685   -8.053 1.13e-14  ***
M5M3_M5        -12.417      1.844   -6.735 6.33e-11  ***
M5M3_M6        -14.058      1.788   -7.863 4.21e-14  ***
M5M4_M4         -9.444      2.030   -4.652 4.59e-06  ***
M5M4_M5        -12.355      1.995   -6.193 1.58e-09  ***
M5M4_M6        -10.021      2.393   -4.188 3.53e-05  ***
---
Signif. codes:  0 '***' 0.001 '**' 0.01 '*' 0.05 '.'

Residual standard error: 5.045 on 368 degrees of freedom
Multiple R-squared: 0.963, Adjusted R-squared: 0.9603
F-statistic: 354.5 on 27 and 368 DF, p-value: < 2.2e-16
```

Nothing much is really happening until we get to the last marker—then we get *stars* everywhere. For the first four markers only a few genotypes are marginally significant and none of them would make it past a *p*-value cutoff of 0.05. They also have reasonably small effects, the largest one being around *6 kg*, the standard deviation is *25 kg*, so just around *0.2* of a standard deviation). Our first four markers were not very exciting. But how do we interpret the results from the fifth marker? *Sex* is still significant and adds around *46 kg* to the weight of males. The sires are different between each other, we hope to have captured the polygenic effects reasonably well. Recall that the differences are shown in relation to *sire 1* and most sires are significantly different from it but not necessarily different between each other. Note that the estimates of sire effects change when different markers are fitted but are still reasonably consistent (the standard errors can help with confidence of estimates).

As for the markers our baseline genotype is *M1_M1*, homozygote for allele *M1*. There are in total 18 genotypes in our data for this marker. We see that the first 5 were not significant and the other 12 were (17 because one is the baseline contrast). If we look closely at the first 5 we will notice that all of them have an allele *M1* in the genotype, but the others do not. And that's our result: allele *M1* has a significant

effect on our trait. What else can we say? Allele *M1* is dominant—remember that
the baseline is *M1_M1* and we are getting almost the same estimates for one or
two "*doses*" of allele *M1*. The allele cannot be recessive because then the difference
between the homozygote *M1_M1* and the heterozygotes would have to be similar to
the difference between having or not allele *M1* (similar to the other 12 values). Could
it be additive? Again we would expect a consistent intermediate value between 0,
1 and 2 copies of the allele—which we do not see here; if anything the estimates
for the heterozygotes seem positive. Of course there might be different interactions
between allele *M1* and each of the other alleles, but that's another story… The mean
of the estimates for genotypes that do not carry allele *M1* is -13.52 kg; thus animals
with allele *M1* will be almost *14 kg* heavier than other animals.

**Our grand conclusion: the candidate gene in linkage with marker 5 is a
potential *suspect* for having an effect and influencing *weight* in cattle. Animals
which carry allele *M1* from marker 5 should show an additional weight gain
of around *14 kg* in relation to other animals. The favorable allele shows a
dominance effect.**

That's the end of this analysis. Did it actually work? I would say yes. The
simulation was based on a mean weight for males of 350 kg (sd $= 10$), females
300 kg (sd $= 10$) and only marker 5, allele *M1* had an effect of 15 kg (sd $= 2$) and
the allele was set up to be dominant.

Some things to keep in mind. This exercise was highly guided—the parameterization worked well. In real life it's not common that the baseline will work so
nicely. First, ignore *p*-values and focus on estimates of the effects, that's much
more meaningful and you will be able to pick the relationships between levels
right away. The main value of the *p*-values is to evaluate your confidence in the
estimate of effects given the data. Also, even if an effect is significant, it might be so
small as to have little value anyhow (and the opposite is true as well). If you don't
like treatment–contrasts you can use a global mean and deviate everything from it.
A good way to view the data is with the simple *plot.design* (Fig. 2.10). Try also the
effects package for nice plots or the *model.tables* function.

```
> plot.design(weight~sex+sire+M5,data=markers,col=
  "blue")
```

Of course, before we dig into the actual estimates as we did here, the easiest way
to see if the marker itself has an effect on the trait is with an anova table. We can
right away see which markers we want to zoom in on.

```
> anova(lm(weight~sex+sire+M1,data=markers))
```

```
Analysis of Variance Table
```

```
Response: weight
          Df Sum Sq Mean Sq   F value Pr(>F)
sex        1 204821  204821 2971.0104 <2e-16 ***
sire       9  21807    2423   35.1467 <2e-16 ***
M1        17    958      56    0.8175 0.6729
```

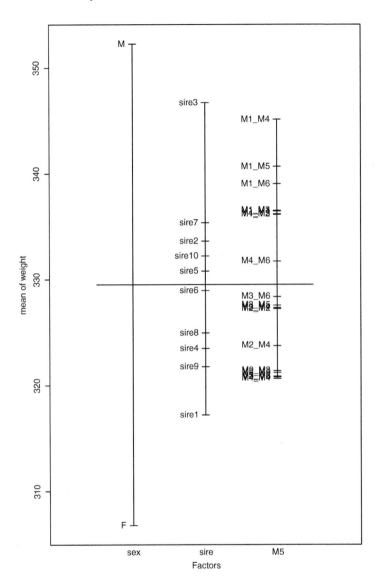

Fig. 2.10 Effect sizes plot of *weight* for factors *sex, sire* and *marker5*

```
Residuals 368   25370        69
---
Signif. codes:   0 '***' 0.001 '**' 0.01 '*' 0.05 '.'
> anova(lm(weight~sex+sire+M2,data=markers))
Analysis of Variance Table
```

```
Response: weight
            Df Sum Sq Mean Sq   F value Pr(>F)
sex          1 204821  204821 2984.1111 <2e-16 ***
sire         9  21807    2423   35.3017 <2e-16 ***
M2          17   1069      63    0.9166 0.5545
Residuals 368  25258      69
---
Signif. codes:  0 '***' 0.001 '**' 0.01 '*' 0.05 '.'

> anova(lm(weight~sex+sire+M3,data=markers))

Analysis of Variance Table

Response: weight
            Df Sum Sq Mean Sq   F value Pr(>F)
sex          1 204821  204821 2971.0585 <2e-16 ***
sire         9  21807    2423   35.1473 <2e-16 ***
M3          17    959      56    0.8179 0.6725
Residuals 368  25369      69
---
Signif. codes:  0 '***' 0.001 '**' 0.01 '*' 0.05 '.'

> anova(lm(weight~sex+sire+M4,data=markers))

Analysis of Variance Table

Response: weight
            Df Sum Sq Mean Sq   F value Pr(>F)
sex          1 204821  204821 3027.2435 <2e-16 ***
sire         9  21807    2423   35.8119 <2e-16 ***
M4          17   1429      84    1.2427 0.2282
Residuals 368  24899      68
---
Signif. codes:  0 '***' 0.001 '**' 0.01 '*' 0.05 '.'

> anova(lm(weight~sex+sire+M5,data=markers))

Analysis of Variance Table

Response: weight
            Df Sum Sq Mean Sq F value    Pr(>F)
sex          1 204821  204821 8047.286 < 2.2e-16 ***
sire         9  21807    2423   95.198 < 2.2e-16 ***
M5          17  16962     998   39.200 < 2.2e-16 ***
Residuals 368   9366      25
---
Signif. codes:  0 '***' 0.001 '**' 0.01 '*' 0.05 '.'
```

Yes, the only marker worth while looking at is marker 5. A note of caution when using *anova*. Terms are tested in the order of the model. Rather crudely this means that after fitting *sex* and *sire*, the marker is tested on the remaining variance not captured by the first two terms. A bit more formally it is the sequential sums of squares (SS)—type I; i.e., the improvement in the error SS as an additional term is added. The order we used here is appropriate and will ensure that the marker does not "*suck up*" variation from the other two terms (and we are really only interested in the marker effects anyhow). If markers were fitted first in the model, they would all appear to be significant. Keep in mind that the order of terms will change the results. To get type II errors (hierarchical) or type III (marginal), the *Anova* function from the library *car* can be used. Type III will give the same results as running *anova* multiple times with a different order for the factors and only looking at the last level each time. One point about *Anova* is that we can test all markers at the same time:

```
> library(car)
> Anova(lm(weight~sex+sire+M1+M2+M3+M4+M5,
+     data=markers), type=3)

Anova Table (Type III tests)

Response: weight
              Sum Sq  Df   F value  Pr(>F)
(Intercept)   143615   1 5574.5479  <2e-16 ***
sex           162947   1 6324.9082  <2e-16 ***.
sire            7529   9   32.4717  <2e-16 ***
M1               335  17    0.7652  0.7328
M2               293  17    0.6699  0.8320
M3               347  17    0.7913  0.7031
M4               574  17    1.3095  0.1845
M5             14151  17   32.3113  <2e-16 ***
Residuals       7729 300
---
Signif. codes:  0 '***' 0.001 '**' 0.01 '*' 0.05 '.'
```

Results are similar but *F*-values are slightly lower, which is not surprising given that more terms were added to the model.

What we did was quite simple. You might want to use a linear mixed effects model (fixed and random effects)—have a look at *lmer* in package *lme4* or *lme* in package *nlme*. For non linear mixed effects models use *nlme* also in package *nlme*. For generalized linear models *glm* is a good option. For those familiar with ASReml there's an R version—but it's not free. For binary data use a logistic regression such as *PLR* from package *MCRestimate* or even a simple χ-square test *chisq.test*.

To end this chapter let's compare the results using our linear model with only fixed effects with a mixed effects model with *sire* as random.

```
> library(nlme)
> linear=coefficients(lm(weight~sex+sire+M5,
+       data=markers))[c(1,2,12:28)]
> random=fixef(lme(weight~sex+M5,random=~1|sire,
+    data=markers))
> linear=data.frame(effect=names(linear),fixed=linear)
> random=data.frame(effect=names(random),random=random)
> comparison=merge(linear,random,by="effect")
> comparison=data.frame(comparison,
+       difference=comparison$fixed-comparison$random)
> print(comparison)
```

	effect	fixed	random	difference
1	(Intercept)	305.860128	315.115009	-9.254880841
2	M5M1_M2	1.504781	1.490549	0.014232509
3	M5M1_M3	2.328878	2.314576	0.014301890
4	M5M1_M4	3.404947	3.453365	-0.048418026
5	M5M1_M5	1.928433	1.971879	-0.043446586
6	M5M1_M6	2.232493	2.254357	-0.021864058
7	M5M2_M2	-16.270931	-16.370461	0.099530184
8	M5M2_M3	-14.598117	-14.625505	0.027387920
9	M5M2_M4	-14.060934	-14.107196	0.046262575
10	M5M2_M5	-14.298566	-14.344659	0.046093036
11	M5M2_M6	-17.000394	-17.034139	0.033744340
12	M5M3_M3	-14.143775	-14.170504	0.026729166
13	M5M3_M4	-13.574197	-13.602779	0.028581779
14	M5M3_M5	-12.417066	-12.440205	0.023139180
15	M5M3_M6	-14.057730	-14.088632	0.030902255
16	M5M4_M4	-9.443498	-9.441622	-0.001875957
17	M5M4_M5	-12.355493	-12.290890	-0.064603120
18	M5M4_M6	-10.020663	-10.080629	0.059965642
19	sexM	46.151039	46.159111	-0.008071354

The differences are small in this example. In general you might expect smaller effects with a random model. What we did above was to extract the coefficients (the ones we wanted to compare—note the use of the index) in *linear* and the fixed effects from our random model in *random*. Then we transformed the results into data.frames with a column called *effect* which we used to combine the two (the function *merge* can be used to combine data.frames based on the contents of a column). You could also try a mixed model with all markers as a random effect (better to use *lmer* for this).

2.6   Useful R Books and Packages

- *Applied Statistical Genetics with R: For Population-based Association Studies* [35]. A good read with a focus on human studies and more emphasis on discrete traits.
- *Statistical Genetics of Quantitative Traits: Linkage, Maps and QTL* [120] and *The Statistics of Gene Mapping* [102] are two good books about mapping and QTL detection with examples in R. Both provide good theoretical foundations.
- *nlme, lme4, mgcv, MCRestimate* for model fitting.
- *lattice* for trellis graphs, ideal for multivariate plots.
- *GeneticsBase* good for storing and handling genetic data, can save a lot of work but it's not very flexible.
- *genetics* excellent generic package, was replaced by *GeneticsBase* but I still prefer this old package.
- *haplo.stats* and *ldDesign* some nice functions to compute power and sample sizes.

Chapter 3
Genome Wide Association Studies

In this chapter we will discuss genome wide association studies (GWAS) using SNP. GWAS present some challenges for biostatistics and bioinformatics—the sheer dimensionality of the data can create storage/retrieval and analysis problems. Quality control and data preprocessing are also important steps in GWAS. We will initially discuss basic database usage for data storage and handling and the main metrics for evaluating the quality of genotypes followed by how to perform a GWAS, multiple testing issues and how to visualize results.

3.1 From Microsatellites and Linkage Analysis to SNP and Genome Wide Association Studies

Until 2004 microsatellites were the most widely used type of DNA markers used in QTL mapping studies in livestock. A relatively large number of microsatellites were known for various species (roughly in the range of 1,000–10,000 depending on the species) and their genomic location was also reasonably well characterized. Note that this predates modern sequencing platforms and assemblies for most species were not yet available.

In general terms, the markers are used to identify regions harboring quantitative trait loci (QTL). In a typical genome scan, about 200–400 markers are genotyped across the genome, and the objective is to identify putative QTL segregating within marker intervals (interval mapping) that are associated with a trait of interest. The limitation of these studies is that the association holds within a family or within a designed cross of divergent breeds but it is generally not valid across different families. The reason is that the markers are generally too far from the QTL, e.g.,

Electronic supplementary material The online version of this chapter (doi: 10.1007/ 978-3-319-14475-7_3) contains supplementary material, which is available to authorized users.

© Springer International Publishing Switzerland 2015
C. Gondro, *Primer to Analysis of Genomic Data Using R*, Use R!,
DOI 10.1007/978-3-319-14475-7_3

the average distance between adjacent markers would be $3,500/200 = 17.5$ cM. A 10 cM distance means that there will be a recombination in 10% of cases. The phase between alleles on the marker and QTL can be different across family groups. Also, the experimental design requirements and the linkage analysis itself can be quite cumbersome in these studies.

Once a broad QTL region is identified, the region can be narrowed down by adding additional new markers (other microsatellites or more often SNP) within the putative QTL region to identify candidate genes and, ultimately, find the functional mutation. If the actual functional mutation can be identified, it can be used directly and linkage is not an issue anymore. This is however a costly exercise and has met limited success apart from some well-known cases of genes with large effects on phenotypes, e.g., myostatin (muscling) and DGAT (milk) mutations.

3.1.1 Single Nucleotide Polymorphism

A new direction was initiated in the early 2000s with sequencing of whole genomes. A by-product of sequencing was the notion that DNA base pairs (=nucleotides) are usually the same in individuals from a species, but they vary about 1 in every 500–1,000. Hence, in a part of the genome, some individuals will have ...AA**C**TGTA ...whereas some others have ...AA**T**TGTA... The variants are called Single Nucleotide Polymorphisms (SNP).

SNP are important because they are all over the genome and at much higher density than microsatellites. A ballpark figure is 20 million reasonably common SNP in a species. Furthermore, many functional mutations will actually be SNP. SNP also allow for fast, accurate, and cheap genotyping. For example, the commercially available ovine SNP chip allows genotyping one individual for more than 50,000 SNP for around $75.00 per animal and prices are dropping rapidly.

3.1.2 Genome Wide Association Studies

The high density of SNP has important implications for association studies. Firstly, QTL experiments do not need to be based on, e.g., large half-sib families or divergent line crosses as we do not need such a rigorous structure to ensure LD. LD can be assumed across families in a population. This will have an added advantage that estimated effects are more likely to be valid across families. Finally, by simultaneously fitting all markers, a statistical model can accommodate the joint effects of all QTL affecting a quantitative trait (rather than a single QTL). Such models are fitted in genomic prediction studies and the combined effect of the SNP can account for a reasonable proportion of the genetic variance observed in a trait. If the marker information is treated appropriately (e.g., snpBLUP) the estimation procedures do not suffer from an excess of false positive results due to multiple testing.

Results from simulation and empirical studies have shown that fitting thousands of gene markers to explain phenotypes for as little as 1,000 subjects is feasible. They have also shown that reasonable power can be obtained if the number of subjects is between 2,000 and 8,000 (depending on trait heritability and effective population size). The broad consensus is that about 60,000 markers are needed for accurate predictions of breeding values (otherwise the distance between markers will be too wide), and even higher numbers/densities are needed to expect successful predictions across populations with different genetic backgrounds. The latter also requires QTL to have similar effects across these populations with results to date suggesting that predictions do not hold well across populations.

We have discussed GWAS with a rather unsubtle bias toward livestock. A nice and much more "*humane*" paper with a historical perspective of genome wide association studies is given by Kruglyak [63]. A good methodological summary on GWAS for human studies is the paper by Bush and More [13]. For a crop plants perspective see Huang and Han [55].

3.2   Experimental Design

We will not go into any in-depth details of design issues. As with any experiment, the design can *make or break* the project. In the previous chapter we discussed some general design concepts such as population wide studies, case–control studies, and family-based studies. Practical considerations unfortunately cannot be avoided and there will have to be some compromise due to availability of funding and samples. A nice discussion of design issues is given by Spencer et al. [106]—there's even an R package developed by the authors that will tell you the power of your experiment given your budget. A book chapter by Roderick Ball has some with nice R examples and an accompanying R package [10]. See also [109] for a discussion of design issues or [62] on power analysis considerations.

A few things to keep in mind though:

- *Sample size.* Sample size and power calculations can give you insights into what you will be able to pick up or not. But the short version is to aim for the largest sample size that your money will buy and you can phenotype.
- *Multiple testing.* Even with a small 10K chip you would still expect 100 false positives for a significance threshold 0.01. And if we use the human 1000K, well... Further on we will discuss multiple testing correction approaches. For the time being just remember that GWAS can yield a large number of spurious results.
- *Rare alleles.* If the trait you are interested in is associated with a rare allele in the population you will need much larger numbers to identify effects in these SNP.
- *Coverage.* In GWAS hopefully there will always be a SNP that is in full (or almost) linkage disequilibrium with the causative gene. But this may not always be the case. If SNP and QTL are not always on the same phase you will lose

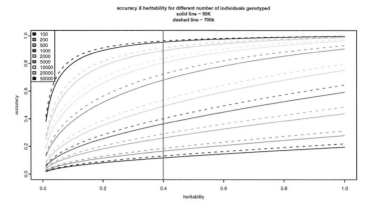

Fig. 3.1 Accuracy × heritability for different number of individuals genotyped on 50k and 700k SNP arrays

some power. Consider using haplotypes for the analysis or at least quantifying recombination if your data will allow it. The population structure will also play a role here. In humans more LD is observed in Caucasian than African populations. A chip suitable for one population might not be adequate for the other in terms of coverage—there can be *ascertainment bias* depending on the data used to select the SNP. In livestock we expect to need less dense arrays than in humans since these populations have smaller effective population sizes and longer stretches of DNA in LD.

- *Size of effect.* It's easier to pick up large effects explained by a few loci. If the trait is determined by many loci of small effects the sample size will have to be considerably larger.

For illustration purposes, Fig. 3.1 shows theoretical accuracies of prediction (this is for genomic selection/prediction, not GWAS) for heritabilities ranging between 0 and 1 and for different number of individuals (between 100 and 50,000) genotyped on 50k and 700k arrays. Note how the 14-fold increase in the number of SNP only increases accuracies between 2 and 10%. The assumptions here were an effective population size of 100 individuals and a genome of 30M [41]. The vertical red line shows changes in accuracy for a trait with heritability 0.4.

3.3 Platforms

It was the availability of chip-based, parallelized SNP assaying of up to one million or even more SNP per individual that allowed genome wide association studies to become so ubiquitous. The main players who produce these arrays for SNP chip genotyping are Affymetrix and Illumina; both producing SNP chips capable of genotyping in excess of one million SNP [79]. Arrays such as the Affymetrix

Human SNP Array 6.0 feature not only SNP (around 900,000) but also copy number polymorphisms (an additional 950,000 probes).The most comprehensive SNP chips are for humans; however Illumina produces, e.g., a 1720k chip for canines, 700k chips for both bovines and sheep as well as a large number of lower density arrays (e.g., 50k bovine, 50k sheep, 60k porcine). Both Illumina and Affymetrix also offer customized SNP chips. Check the manufacturers websites for detailed documentation of the technologies in use and the various *off the shelf* chips available.

As the sequencing technology rapidly evolves, we can expect that array-based genotyping will be replaced by cheap sequencing methods in the very near future (*GBS*—genotyping by sequencing). With sequence-based data there are differences in the preprocessing and quality control steps but the analysis itself is still largely the same.

3.4 Preprocessing and Quality Control

In the previous chapter we spent quite some time cleaning up the data for the analysis. GWAS also demand a fair amount of preprocessing and quality control. We will not discuss the phenotypes here—we went over some common steps in Chap. 2. Here we will focus only on the SNP data.

We will work with a small sample from a real dataset from sheep. The platform is an Illumina SNP chip with 54,977 SNP and the data file was generated by Genome-Studio (proprietary software from Illumina). We have 83 animals genotyped, not nearly enough to even get started in a real association study but enough to discuss the principles.

The next sections on data handling and quality control are largely based on the book chapter [43] and reproduced here with kind permission from the publisher. A fully automated analysis pipeline in R that performs QC and stores data is described in [45].

3.4.1 Storing and Handling Data

This dataset can quite easily be handled directly in R, but for larger datasets dimensionality can become a problem—the whole dataset will not fit into the memory of common desktop machines. The simplest workaround is to store the data in a database and retrieve only the parts of the data that are needed at any given time. R can interface quite easily with databases—SQL queries can be sent straight from R to the database and retrieved records stored as a data.frame.

Databases are by far the best approach to manage large datasets. While for large collaborative projects it is essentially mandatory to have a dedicated database manager to design the database and manage data storage/access, it is still easy to implement robust solutions for smaller projects and worthwhile to have a general understanding of their usage.

SNP array data will usually come in two formats: a proprietary database structure developed by the chip manufacturer or a flat file. We will work only with the flat files. Our first step is to build a database from the flat files for further downstream analysis in R. We will build the database straight from R, but of course you don't need it for this. There are many options for working with databases. We will use *SQLite*. The key advantages are that it connects well to R—the annotation packages that we will see in the next chapters are all built with it; there's no installation involved—a single 500 kb executable is all you need; databases can simply be copied across machines without any further installation; to access a database from R all we need is the *RSQLite* package.

For our example we will use two files: a genotypes file and a map file. The first (Fig. 3.2) contains all genotype calls for all SNP and all samples; the second holds mapping information of the SNP, e.g., chromosome, physical location, etc.

For our database we need to create a schema (a schema is simply a text file that describes the database structure and is used to create the initial empty DB) with the tables and columns we want in it. The map file (the simplified version used here) has only three columns: SNP identifier, chromosome, and position. The data file (Fig. 3.2) has seven columns: SNP identifier, sample identifier, allele 1, allele 2, X, Y, and GC score (we will discuss the last three later on). As mentioned before, this data file was generated by GenomeStudio. There are many other columns that can be added/removed to the data—keep in mind that the actual columns and their order can change between datasets. Also note that GenomeStudio has extensive

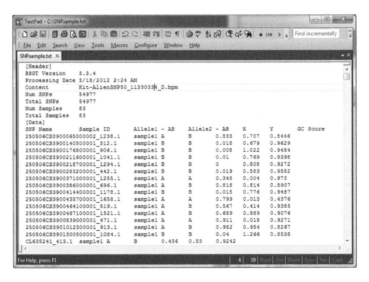

Fig. 3.2 Screenshot of SNP chip data

capability for data filtering as well, and strictly speaking most of what we are doing here could be done straight in GenomeStudio (albeit at a rather high price tag and with less flexibility).

Back to the data, notice that we have some *nuisance* information lines—9 in this case. A simple schema will consist of two tables, one for each of the files and one column for each source of information. A *snpmap* table with SNP identifier, chromosome and position, and a data table with the seven columns from our dataset. The schema could look like

```
CREATE TABLE snpmap(
  name,
  chromosome,
  position
);

CREATE TABLE SNP(
  snp,
  animal,
  allele1,
  allele2,
  x,
  y,
  gcscore
);

CREATE INDEX snp_idx ON SNP(snp);
CREATE INDEX animal_idx ON SNP(animal);

CREATE INDEX chromosome_idx ON snpmap(chromosome);
```

Note that the above is **not** R code. It is simply a plain text file with the structure (description) of the DB that we want to create (you can write and save this using, e.g., *Notepad*). The table and columns names are discretionary; here we decided to hold mapping information (three columns) in *snpmap* and the genotypes in *SNP* (seven columns). Note that indices (*CREATE INDEX*) are created at the end to make sure that searches by SNP ID, sample ID, or chromosome are fast to execute (indices make the database slow to build but speed up the queries dramatically). A good source of information for SQLite syntax can be found at http://www.sqlite.org/. Once we've saved our schema in a text file (e.g., as snpDB.sql or schema.txt) we are ready to create the database and populate it with the data. This can be done from the command line on the console (also **not** in R) provided, of course, *sqlite* is installed and in the OS path with:

```
sqlite SNPsmall < snpDB.sql
```

and this creates a new database *SNPsmall* with the previous schema ready to be populated with the genotype data. Excel users might find it useful to think of a schema as a text file that defines the names and number of *sheets* in a spreadsheet and within each *sheet* (table), there's a certain number of column headers.

But this is really just to briefly illustrate the internal structure of a database. In practice it is quite simple to create a new database directly in R, and for our example not even the schema is needed. First open R and load the *RSQLite* package

```
> library(RSQLite)
```

Then run the following code:

```
> con=dbConnect(dbDriver("SQLite"),
+ dbname = "chapter3/SNPsmall")

> dbWriteTable(con,"snpmap","chapter3/SNPmap.txt",
+  header= TRUE, append=T, sep="\t")

> dbWriteTable(con,"SNP","chapter3/SNPsample.txt",
+ append=T, header=TRUE, skip=9,sep="\t")
```

Let's go over the code. The first line simply creates a new blank database called *SNPsmall* using the SQLite database driver (R has DBMS's for most common database engines). In the same line a connection (function *dbConnect*) to the new database is created (if the database already existed, R would simply open a connection to it).

Now we have an empty DB similar to what we did using straight *sqlite* but without any tables/fields information. R can simply create tables and fields directly from the flat files themselves. To populate the DB (once connected to it using *dbConnect*) we can upload our flat files of genomic data *SNPmap.txt* and *SNPsample.txt* using the function *dbWriteTable* (once for each file). The function takes quite a few arguments: the database connection *con* that was created in the previous line, the name of the a new table to be created in the DB (*snpmap* and *SNP*), the name of the flat file to import, if there is a header in the file it can be used to create the field/column names, if appending or not to the database, the separator used between columns in the data and how many lines to skip if there are extraneous header files. Note that in this example the top nine lines from the genotypes file had to be removed (Fig. 3.2). Note: a source of angst when creating databases in R is what is used as *end of line (eol)* in the file. While R automatically recognizes \n (Linux) or \r\n (Windows), SQLite does not. An additional parameter to *dbWriteTable* is *eol* (e.g., eol=")"\r\n").

We can have a look at the tables and fields (columns) in the DB using *dbListTables* and *dbListFields*

```
> con=dbConnect(dbDriver("SQLite"),
+     dbname = "chapter3/SNPsmall")
> dbListTables(con)
```

```
[1] "snpmap" "SNP"
```

```
> dbListFields(con,"snpmap")
```

```
[1] "name"        "chromosome" "position"
```

```
> dbListFields(con,"SNP")
```

```
[1] "snp"        "animal"    "allele1" "allele2" "x"
[6] "y"          "gcscore"
```

The function *dbGetQuery* is used to send an SQL query to the DB and return the data in a single step. A two step approach is using *dbSendQuery* and *fetch*, but we will not discuss these here. The syntax for *dbGetQuery* is *dbGetQuery(connection name,"SQLquery")*. For example, if we want to retrieve the number of records in a table

```
> dbGetQuery(con,"select count (*) from snpmap")
```

```
  count (*)
1      54977
```

```
> dbGetQuery(con,"select count (*) from SNP")
```

```
  count (*)
1    4563091
```

That looks right. There are 54,977 records in *snpmap* and we know our chip has 54,977 SNP so that matched up well. The number of records in *SNP* is also fine—it should be the number of samples times the number of SNP

```
> 54977*83
```

```
[1] 4563091
```

If we want to retrieve sample ids we would do something as

```
> animids=dbGetQuery(con,
+       "select distinct animal from SNP")
> dim(animids)
```

```
[1] 83  1
```

```
> head(animids)
```

```
     animal
1   sample1
2  sample10
3  sample11
4  sample12
5  sample13
6  sample14
```

Herein we will not discuss SQL queries or syntax. Any general SQL book will cover most of the common needs (see for example [91]). All we really need to know is how to use *select * from tableName where columnName= "mysearch"*. For example let's retrieve all data associated with the first sample.

```
> animids=as.vector(animids$animal)
> hold=dbGetQuery(con,paste(
+       "select * from SNP where animal='",animids[1],"'",
+       sep=""))
> dim(hold)

[1] 54977      7

> head(hold)
```

	snp	animal	allele1	allele2
1	250506CS3900065000002_1238.1	sample1	A	B
2	250506CS3900140500001_312.1	sample1	B	B
3	250506CS3900176800001_906.1	sample1	B	B
4	250506CS3900211600001_1041.1	sample1	B	B
5	250506CS3900218700001_1294.1	sample1	B	B
6	250506CS3900283200001_442.1	sample1	B	B

	x	y	gcscore
1	0.833	0.707	0.8446
2	0.018	0.679	0.9629
3	0.008	1.022	0.9484
4	0.010	0.769	0.9398
5	0.000	0.808	0.9272
6	0.019	0.583	0.9552

In the first line we just changed the data.frame with animal ids to a vector—saves some indexing work. Then we retrieved the data for the first sample from our vector of animal ids. It's quite easy to picture a loop for each animal—read in one animal, run some analysis or other, read in the next animal... Notice the use of *paste* to create a query string and also the rather awkward use of single and double quotes— we need quotes for the R string and we also need to include a single quote for the SQL query in the DB. Just the query string looks like

```
> paste("select * from SNP where animal='",animids[1],
"'",+ sep="")

[1] "select * from SNP where animal='sample1'"
```

We already have a vector for the samples. Let's also get a vector of SNP.

```
> snpids=as.vector(dbGetQuery(con,
+       "select distinct name from snpmap")[,1])
> length(snpids)
```

```
[1]  54977
```

```
> head(snpids)
```

```
[1]  "250506CS3900065000002_1238.1"
[2]  "250506CS3900140500001_312.1"
[3]  "250506CS3900176800001_906.1"
[4]  "250506CS3900211600001_1041.1"
[5]  "250506CS3900218700001_1294.1"
[6]  "250506CS3900283200001_442.1"
```

So, we managed to successfully create a database, connect to it, add the data from the flat files, and retrieve data from it into R. Before we close this section, recall that in our schema we created indexes to make data retrieval faster (you would have noticed that it is somewhat slow right now). We can also add indexes to our database straight from R with

```
>dbGetQuery(con,
+ "CREATE INDEX chromosome_idx ON snpmap(chromosome)")
```

```
>dbGetQuery(con,"CREATE INDEX snp_idx ON SNP(animal)")
>dbGetQuery(con,"CREATE INDEX ID_idx ON SNP(snp)")
```

This is just sending an SQL command to the database (here, create the indexes). Any valid SQL syntax can be sent to the DB as a string (meaning that anything that can be done in SQL can be done from R). A significant performance improvement is noticeable if you again retrieve the SNP from the database

```
> snpids=as.vector(dbGetQuery(con,
+       "select distinct name from snpmap")[,1])
```

And one last thing. When we are finished with the DB we should close the connection

```
> dbDisconnect(con)
```

```
[1]  TRUE
```

3.4.2   Quality Control

Now we are ready to do some quality control on our data. Various metrics are commonly used for QC in GWAS. There is still some level of subjectivity in these, particularly when setting thresholds. The statistics are performed either across SNP or across samples, and the objective is to remove bad SNP or samples because the genotypes are incorrect (or unreliable). The main point with QC is that quite a few of the filtering parameters used are based on some sort of population metric and the definition of good/bad is based on a *deviation* in relation to other SNP and/or samples. The challenge is to identify artificial variation introduced due to errors from true population variation.

To illustrate this point, genotypes tend to be susceptible to ascertainment bias which is caused by the SNP discovery process being based on a relatively small number of individuals. This bias can complicate estimates of genetic diversity and affect call rates [5, 64]. This effect is even stronger when an array designed for, e.g., a particular breed is used for genotyping a more genetically distant one (e.g., *Bos taurus* and *Bos indicus*). In these scenarios, the viability of the platform, has to be evaluated [93] prior to wide adoption and various filtering criteria thresholds have to be tested.

Typical SNP filtering criteria include percent genotyping fail, median call rates, genotyping quality scores (GC score), minor allele frequencies and SNP that are not segregating, deviation of heterozygosity in number of standard deviations, and deviation from Hardy–Weinberg equilibrium. Sample filtering criteria include call rates, deviation of heterozygosity in number of standard deviations, and correlation between samples. Individual chromosomes sometimes are also selected for exclusion (e.g., sex chromosomes).

Let's start with across SNP analyses.

3.4.2.1 Genotype Calling and Signal Intensities

SNP alleles are usually coded as A/B in Illumina (as in our example—Fig. 3.2) or the actual nucleotides are used. This changes depending on the platform, laboratory or the export parameters used to generate the genotype file. Preference should be to use a simple reference for alleles and an additional DB table with all pertinent information for the SNP. Let's have a look at the first SNP (snpids[1]) in the dataset.

```
> con=dbConnect(dbDriver("SQLite"),
+       dbname = "chapter3/SNPsmall")
> snp=dbGetQuery(con,
+       paste("select * from SNP where snp='",
+       snpids[1],"'",sep=""))
> dim(snp)

[1] 83   7

> head(snp)

                             snp   animal allele1 allele2
1 250506CS3900065000002_1238.1 sample1       A       B
2 250506CS3900065000002_1238.1 sample5       A       B
3 250506CS3900065000002_1238.1 sample6       A       B
4 250506CS3900065000002_1238.1 sample7       B       B
5 250506CS3900065000002_1238.1 sample8       B       B
```

```
6  250506CS3900065000002_1238.1 sample9            B        B
          x        y gcscore
1 0.833 0.707    0.8446
2 0.829 0.714    0.8446
3 0.816 0.730    0.8446
4 0.031 1.132    0.8446
5 0.036 1.146    0.8446
6 0.037 1.150    0.8446

> snp$allele1=factor(snp$allele1)
> snp$allele2=factor(snp$allele2)
> summary(snp$allele1)

 A  B
36 47

> summary(snp$allele2)

 A  B
 6 77
```

We first restore the DB connection that we had closed then send an SQL query to retrieve data for the first SNP. Convert alleles into factors (usually data is returned as character) and then summarize the allele information. There are no missing values in our data and as we expected, there are only two alleles—*A* and *B*. If there were missing values (missing calls, to use the terminology) we would see a third factor (e.g., *NA* or "-"). Our data also has an *X* and a *Y* column (these are specific to Illumina). These are the normalized intensities of the reads for each of the two alleles. Allele calls are assigned based on the signal intensity of the fluorescence read by the scanner. These intensities can be plotted as an *XY* plot. We would expect that one of the homozygous genotypes would show high *X* values and low *Y* values while the other homozygote would be the opposite. Heterozygotes would be somewhere between the two. If the technology was 100 % accurate we would have only three points on the plot and all samples would have the same intensity measures; but since there is considerable variation in intensities, what we do observe are three clouds (clusters) of data which will hopefully separate well between each other (Fig. 3.3). To plot the data

```
> snp=data.frame(snp,
+      genotype=factor(paste(snp$allele1,
+      snp$allele2,sep=""),
+      levels=c("AA","AB","BB")))
> plot(snp$x,snp$y,col=snp$genotype,
+      pch=as.numeric(snp$genotype),
+      xlab="x",ylab="y",
+      main=snp$snp[1],cex.main=0.9)
```

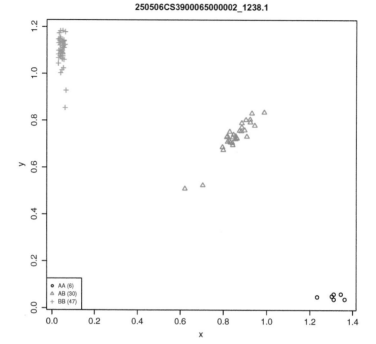

Fig. 3.3 XY plot of intensity reads

```
> legend("bottomleft",paste(levels(snp$genotype),
+     " (",summary(snp$genotype),")",sep=""),
+     col= 1:length(levels(snp$genotype)),
+     pch= 1:length(levels(snp$genotype)),
+     cex=0.7)
```

We coded the genotypes by color and symbol to make it easier to distinguish between them. But for this we had to add a new column to our data with the genotypes—that's what we do in the first line of code, and then we plot the *X* and *Y* data using the same plotting functions we discussed before. Notice how the genotypes clearly cluster into three discrete groups—an indication of good data. Of course you do not want to look at each of these plots one by one for each SNP (especially the human 1000K array!). Common practice is to go back to these plots after the association test and make sure that the SNP data looks ok at least for the significant SNP. There are some methods to summarize the clusters into an objective measurement, e.g., sums of the distances to the nearest centroid of each cluster and the individual calls.

Another metric included in the data is the *GC score*, without any in-depth details, it is a measure of how reliable the call is (essentially, distance of the call to the

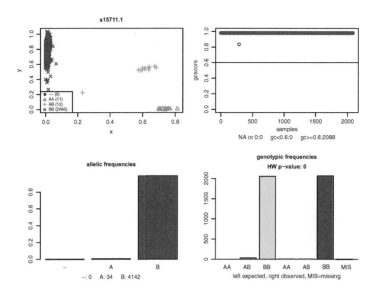

Fig. 3.4 Example of a good quality SNP. *Top left*: clustering for each genotype (non-calls are shown as *black circles*). *Top right*: GC scores. *Bottom left*: non-calls and allelic frequencies (actual counts are shown under the histogram). *Bottom right*: genotypic counts, on the left-hand side the expected counts and on the right the observed counts; the *last block* shows number of non-calls

centroid as we mentioned above) on a scale from 0 to 1. Some labs will not assign a call to *GC scores* under 0.25. Another common *magic number* is to cull reads under 0.6 (and projects working with human data may use even higher thresholds of 0.7–0.8).

```
> length(which(snp$gcscore<0.6))

[1] 0
```

For this SNP all *GC scores* are above 0.6. Figures 3.4 and 3.5 exemplify what a good and a bad SNP look like. We might want to cull individual reads based on a threshold *GC score* value, but we might also remove the whole SNP if, for example, more than 2 or 3 % of the genotyping for the SNP failed or if the median *GC score* is below a certain value (say 0.5 or 0.6). Again, the SNP we are analyzing is fine.

```
> median(snp$gcscore)

[1] 0.8446
```

3.4.2.2 Minor Allele Frequency and Hardy–Weinberg Equilibrium

Population-based metrics are also employed. A simple one is the *minor allele frequency—MAF*. Not all SNP will be polymorphic, some will show only one

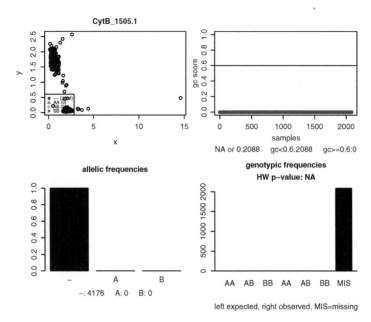

Fig. 3.5 Example of a bad quality SNP. *Top left*: clustering for each genotype (non-calls are shown as *black circles*—here all samples). *Top right*: GC scores. *Bottom left*: non-calls and allelic frequencies (actual counts are shown under the histogram). *Bottom right*: genotypic counts, on the left-hand side the expected counts and on the right the observed counts; the *last block* shows number of non-calls

allele across all samples (monomorphic) or one of the alleles will be at a very low frequency. The association between a phenotype and a rare allele might be supported by only very few individuals (no power to detect the association), in this case the results should be interpreted with caution. To avoid this potential problem, SNP filtering based on MAF is often used to exclude low MAF SNP (usual thresholds are between 1 and 5 %), but it is worthwhile to check the sample sizes and estimate an adequate value for your dataset. To illustrate, at a MAF cutoff value of 5% there would only be 25 homozygous samples in 10,000 (assuming H–W equilibrium); the weightings for the genotypes can be adjusted to minimize these frequency problems but caution should be exercised if it is worthwhile to keep such SNP.

In our example the allelic frequencies for the SNP are

```
> alleles=factor(c(as.character(snp$allele1),
+      as.character(snp$allele2)),
+      levels=c("A","B"))
> summary(alleles)/sum(summary(alleles))*100

      A        B
25.3012 74.6988
```

The frequencies are reasonable, around one-quarter A allele and three-quarters B allele. But again, the point to consider is the objective of the work and the structure of the actual data that was collected. For example, if QC is being performed on mixed samples with an overrepresentation of one group, it is quite easy to have SNP that are not segregating in the larger population but are segregating in the smaller one—the MAF frequency in this case will essentially be the proportion of the minor allele from the smaller population in the overall sample. And if the objective of the study was to characterize genetic diversity between groups, the interesting SNP will have been excluded during the QC stage.

The next metric is *Hardy–Weinberg (HW) equilibrium*. For a quick refresher, the Hardy–Weinberg "*law*", independently proposed by G.H. Hardy and W. Weinberg in 1908, describes the relationship between genotypic frequencies and allelic frequencies and how they remain constant across generations (hence also referred to as Hardy–Weinberg equilibrium) in a population of diploid sexually reproducing organisms under the assumptions of random mating, an infinitely large population and other assumptions.

Consider the bi-allelic SNP with variants A and B at any given locus, there are three possible genotypes: *AA, AB*, and *BB*. Let's call the frequencies for each genotype D, H and R. Under random mating (assumption of independence between events) the probability of a cross $AA \times AA$ is D^2, the probability for $AA \times AB$ is $2DH$ and the probability for $BB \times BB$ is R^2. If p is the frequency of allele A ($p = D + H/2$) then the frequency of B will be $q = 1 - p$ and consequently the genotypic frequencies D, H, and R will respectively be p^2, $2pq$, and q^2.

The relationship model in itself is simply a polynomial expansion (binomial in the above example with two alleles), so it can easily be extended to scenarios with multiple alleles or n-ploid organisms, it can also accommodate inheritance patterns of genes on sex chromosomes and other particularities of individual systems. Whatever the experimental scenario, Hardy–Weinberg equilibrium can be seen as the null hypothesis of the distribution of genetic variation when no biologically significant event is occurring in the population. This in turn allows testing quantitative predictions on empirical data. This "*law*" is, in fact, an informal theorem that depicts the mathematical consequences of Mendelian inheritance at a population level. Rigorously the theorem can be proven within a system that specifies the mechanisms of inheritance and all the other causal modulators. Naturally real populations will not strictly adhere to the assumptions for Hardy–Weinberg equilibrium, but the model is however quite robust to deviations. When empirical observations are in a statistical sense significantly different from the model's predictions, there is a strong indication that some biologically relevant factor is acting on this population or there are genotyping errors in the data. This is where HW becomes controversial—it can be hard to distinguish a genotyping error from a real population effect. Common p-value thresholds for HW are, e.g., 10^{-4} or less (I tend to use multiple testing corrected p-values, so much lower cutoffs).

To calculate HW for our SNP in R

```
> obs=summary(factor(snp$genotype,
+       levels=c("AA","AB","BB")))
> print(obs)

AA AB BB
 6 30 47

> hwal=summary(factor(c(as.character(snp$allele1),
+       as.character(snp$allele2)),levels=c("A","B")))
> hwal=hwal/sum(hwal)
> print(hwal)

        A         B
0.2559524 0.7440476

> exp=c(hwal[1]^2,2*hwal[1]*hwal[2],hwal[2]^2)*sum(obs)
> names(exp)=c("AA","AB","BB")
> # chi-square test
> # with yates correction
> xtot=sum((abs(obs-exp)-c(0.5,1,0.5))^2/exp)
> # get p-value: high chi-square, low pvals
> pval=1-pchisq(xtot,1)
> print(pval)

[1] 0.8897645
```

It's just a typical χ-square test. The only interesting part is the Yates correction used when adding up the χ-square values. Yates correction is appropriate for small sample sizes but can lead to biases in larger datasets. Generally the interpretation of results (which SNP to include/exclude) is the same but actual χ-square values can be quite unexpected. And before we forget, yes—the SNP is in Hardy–Weinberg equilibrium.

3.4.2.3 Quality Control Across Samples

Quality control across samples is quite similar to what we did with the SNP. If 2 or 3% of the genotypes are missing it is probably a good idea to exclude the sample, provided there is nothing particularly different about it (an extreme example: a single sample comes from a different breed). Low median GC scores are also indicative of problems in the DNA extraction or genotyping.

Another criterion that we have not discussed so far is correlation between samples. If samples show very high correlations they might have to be excluded. This could be an indication of wrong genotyping (e.g., samples wrongly duplicated).

However, some studies have used duplicated samples to evaluate reproducibility of genotypes; in this case the correlation can be informative about the quality of the data.

As a ballpark figure, random (simulated) genotypes in HW equilibrium would on average show a correlation of 0 but in real populations this tends to be around 0.3–0.6. Of course, this is highly connected to the structure of your data—a case–control study with random samples is different from a half-sib project in livestock. Correlations are computationally intensive because you need all data stored in memory as a matrix to calculate all pairwise correlations. The R function is *cor(genotypesMatrix)*. If you cannot fit the data matrix into memory consider some roundabouts such as reading samples in pairs for all combinations (can take a long time to run!) or estimate correlations from a random subset of the SNP.

3.4.2.4   Heterozygosity

Another useful metric is heterozygosity, which is simply the proportion of heterozygotes in relation to all genotypes. You can also check heterozygosity on SNP and compare to the expected heterozygosity (or gene diversity), it's just more common to evaluate heterozygosity on the samples. Essentially if a sample's heterozygosity is too high it can be an indication of DNA contamination. Removal of samples ± 3 standard deviations from the mean is reasonable. We will need the heterozygosity for all samples before we can look for outliers. We could just go over all samples in the DB, calculate (and store) the heterozygosity for each subject and discard the data. You might have to do that if the dataset is too large, but since our example is quite small let's build a matrix with the genotype counts (*sumslides*) for all samples and a matrix of *SNP × sample* (*numgeno*)—our entire dataset.

```
> sumslides=matrix(NA,83,4)
> rownames(sumslides)=animids
> colnames(sumslides)=
+     c("-/-","A/A","A/B","B/B")
> # hold reshaped (numeric data)
> numgeno=matrix(9,54977,83)
> for (i in 1:83)
+ {
+    hold=dbGetQuery(con, paste(
+    "select * from SNP where animal='",animids[i],"'",
+    sep=""))
+
+    hold=data.frame(hold,
+      genotype=factor(paste(hold$allele1,hold$allele2,
+      sep=""),levels=c("--","AA","AB","BB")))
+
+    hold=hold[order(hold$snp),]
```

```
+    sumslides[i,]=summary(hold$genotype)
+    temp=hold$genotype
+    levels(temp)=c(9,0,1,2)
+    numgeno[,i]=as.numeric(as.character(temp))
+    # change to 9 genotypes under GC score cutoff
+    numgeno[which(hold$gcscore<0.6),i]=9
+ }
> rownames(numgeno)=hold$snp
> colnames(numgeno)=animids
> dim(sumslides)

[1] 83    4

> dim(numgeno)

[1] 54977     83

> head(sumslides)

            -/-    A/A    A/B    B/B
sample1    838  15818  20100  18221
sample10   777  15397  21367  17436
sample11   803  15381  21564  17229
sample12   763  15440  21145  17629
sample13   822  16257  19524  18374
sample14   750  15637  21014  17576

> numgeno[1:10,1:3]

                                        sample1  sample10  sample11
250506CS3900065000002_1238.1                 1         2         1
250506CS3900140500001_312.1                  2         1         2
250506CS3900176800001_906.1                  2         1         2
250506CS3900211600001_1041.1                 2         2         2
250506CS3900218700001_1294.1                 2         2         1
250506CS3900283200001_442.1                  2         0         1
250506CS3900371000001_1255.1                 0         2         2
250506CS3900386000001_696.1                  1         1         0
250506CS3900414400001_1178.1                 2         2         2
250506CS3900435700001_1658.1                 9         9         9
```

What did we do? We defined a matrix to store genotype counts for each sample (*sumslides*) and gave names to the rows and columns just to make it easier to identify in the output (see above). Notice that we used three genotypes plus −/− for missing genotypes. Another matrix *numgeno* was created to store all genotypic data. Then we made a loop to query the DB and extract data for each animal, sorted the data by SNP (using *order*) to make sure that the data returned by the DB is always in the same order; summarized the genotypic data in their classes and added the results

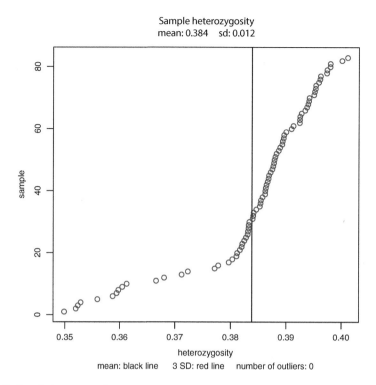

Fig. 3.6 Sample heterozygosity

to *sumslides*. Then we re-leveled the genotypes into numeric format (*9—missing, 0—AA, 1—AB, 2—BB*), we will see why further on. And in the last line of the loop we set all genotypes with *GC scores* under 0.6 as missing. Finally some housekeeping, assign names to the rows and columns so we can identify SNP and samples and check if everything looks alright.

To calculate and plot heterozygosities is quite simple now that we have the data. All we have to do is divide the number of heterozygotes by the total genotypes; calculate the mean and standard deviation, and then calculate the values for *3SD* to each side of the mean. Finally we plot (Fig. 3.6) the data and add lines for the mean and *3SD* (the SD lines do not show up in the plot since we have no outliers in our data). The plot itself is not so simple but it highlights the versatility of R for creating graphs—a simple version is just *plot(sort(samplehetero))*. Finally, we save the figure as a PDF.

```
> samplehetero=sumslides[,3]/(sumslides[,2]+
+       sumslides[,3]+sumslides[,4])
> # outliers 3 SD
> up=mean(samplehetero)+3*sd(samplehetero)
> down=mean(samplehetero)-3*sd(samplehetero)
> hsout=length(which(samplehetero>up))
```

```
> # number of outliers
> hsout=hsout+length(which(samplehetero<down))
> plot(sort(samplehetero),1:83,col="blue",cex.main=0.9,
+       cex.axis=0.8,cex.lab=0.8,
+       ylab="sample",xlab="heterozygosity",
+       main=paste("Sample heterozygosity\nmean:",
+       round(mean(samplehetero),3),"      sd:",
+       round(sd(samplehetero),3)),
+       sub=paste("mean: black line       ",3,
+       "SD: red line      number of outliers:",hsout),
+       cex.sub=0.8)
> abline(v=mean(samplehetero))
> abline(v=mean(samplehetero)-3*sd(samplehetero),col="red")
> abline(v=mean(samplehetero)+3*sd(samplehetero),col="red")
> dev.print(file="images/samplehetero.pdf",
+       device=pdf,width=6,height=6) # save plot
```

We still have not calculated the correlation matrix. The function *cor* will only work with numeric data, hence us changing the genotypes to numeric format (it's also much smaller to store—10M × 208M in the original file).

```
> animcor=cor(numgeno)
> library("gplots")
> hmcol=greenred(256)
> heatmap(animcor,col=hmcol,symm=T,labRow=" ",labCol=" ")
```

In the first line we calculate the correlation matrix—simple Pearson correlation and then plot the results as a heatmap (Fig. 3.7). Heatmaps are excellent to visualize relationships between data and we will revisit them in the next chapters. The library *gplots* has some nice graphing functionalities. The function *greenred* defines the number of colors we want to use in the heatmap; more colors will evidence smaller changes, less colors help bring out higher levels of the data structure. Try to play around with the number of colors to get a feeling for it. The argument *symm = T* in *heatmap* is used to tell the function that the matrix is symmetric.

Note that missing data was replaced by 9—this greatly inflates differences between samples and, on the other hand, strongly pulls together samples with a lot of missing data. Keep in mind that this is for QC purposes only; such an approach should not be used to estimate genomic relationships from the data.

A couple of last comments before we move on to the association tests: (1) what we discussed here was across all SNP and/or samples. With case–control studies it is worthwhile running these QC metrics independently on cases and controls and then checking the results for consistency. (2) We have not plotted any results based on mapping information, it is a good idea to plot, e.g., HW statistics per chromosome to see if there are any evident patterns such as a block on the chromosome that is consistently out of HW. In Appendix A is a full example QC report for the dataset we have been looking at—the entire report is automatically generated using R [45]. A couple of good review papers for quality control issues in GWAS are given by Ziegler et al. [125] and Teo [109].

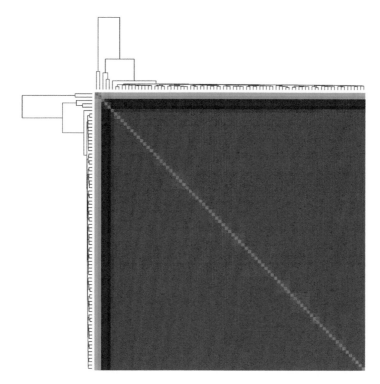

Fig. 3.7 Heatmap of sample correlations

3.5 Single SNP Analysis

Somewhat surprisingly most of the hard work is already done. At least in so far
as simple analyses are concerned—remember, this is a *primer* after all! Our starting
point is that we applied all the above QC metrics discussed so far and built a numeric
matrix of *SNP* × *sample* similar to *numgeno* as we discussed above. The only
difference is that the genotypes of unreliable SNP were replaced with *9* and the same
for unreliable samples. This matrix was saved as a text file called *SNPxSample.txt*.

```
> genotypes=read.table("chapter3/SNPxSample.txt",
+       header=T,sep="\t",na.strings = "9",
+       colClasses = "factor")
> dim(genotypes)

[1] 54977    83
```

We know the data format, so it's much easier to import the file into R in a *ready to
use* format. If we wanted to do linear regressions on our data we would have defined
c("character",rep("factor",83)) (we have to specify the type of each column because
R will return an error if we try to define the row names as numeric). But here we

are going to fit genotypes as factors in an anova. We also defined that our missing value is 9, so no need to fix this up later on. Just out of convenience (and a sense of aesthetics) let's change the names of the factors from *0, 1* and *2* to *AA, AB* and *BB*. Note that *NA* is not within brackets—it's a data type (missing).

```
> for (i in 1:length(genotypes[1,]))
+      levels(genotypes[,i])=c("AA","AB","BB",NA)
```

There are some SNP and/or samples that were discarded (entire row/column set to 9–*NA*). There's little point in analyzing these, so let's remove them from our data.frame. Of course we could have removed them when we did the QC, but in more complex analysis we might want to, e.g., impute the data—it's good to be flexible.

```
> indexsnp=apply(genotypes,1,
+      function(x) length(which(is.na(x)==T)))
> indexsnp=which(indexsnp==length(genotypes[1,]))
> indexsample=apply(genotypes,2,
+      function(x) length(which(is.na(x)==T)))
> indexsample=which(indexsample==length(genotypes[,1]))
> length(indexsample)
```

```
[1]  0
```

```
> length(indexsnp)
```

```
[1]  5458
```

Now here we did something different. The intuitive approach (at least from a programmer's perspective) would be to loop across rows and count the number of *NA*s and then repeat across columns. So far we have been very liberal in the use of loops, but the truth is that R is not good with loops—or at least it is very slow with them. But R has in-built functions that are very effective for recursive operations, one of these is *apply*. What this function does is go over all rows or all columns of a data.frame and repeat the same procedure on each one. The syntax is *apply(mydata.frame,direction,procedure)*. The first argument is the name of the data.frame we want to work on (our *genotypes*); the second argument is the direction—along rows (1) or along columns (2) and the third argument is what we want to repeat (e.g., calculate means or standard deviations). Here all we wanted was to get the number of SNP or samples which are missing (*NA*). A bit harder to follow is what *function(x)* is all about. This is quite similar to a routine in a programming language. We can use *function* to encapsulate a series of commands, it will take in declared arguments (e.g., variable1, variable2, etc.), execute the operations and return a result of these. This is what we did above. We declared a *function* that takes in data *x* which is one of the rows or columns of the data.frame, calculates the number of *NA*s in that row or column and returns it to the caller.

So, what *apply* does is hmm, *apply* a function over either the rows or columns of a data.frame and automatically concatenate the results into the most logical data structure (in our example *apply* returns a vector of integers).

Of course functions can be used out of *apply*. In reality, each time we use a command of the type *commandname(mydata)* we are using a function. An R package is just a library of functions. We can make our own functions, for example

```
> multiplier = function(x,y,z)
+ {
+              res=x*y*z
+              return(res)
+ }
> multiplier(5,9,13.4)

[1]  603
```

We created a function called *multiplier* which takes in three arguments, multiplies them and returns the multiplied value. Now, whenever we need to multiply three values we can use this function. Hint: many functions in R if you type in the name without an argument will return the code used in the function—quite handy to see what's behind the curtain and to learn a bit more about R without having to delve into the source code.

```
> var

function (x, y = NULL, na.rm = FALSE, use)
{
    if (missing(use))
        use <- if (na.rm)
            "na.or.complete"
        else "everything"
    na.method <- pmatch(use,c("all.obs", "complete.obs",
        "pairwise.complete.obs","everything",
        "na.or.complete"))
    if (is.na(na.method))
        stop("invalid 'use' argument")
    if (is.data.frame(x))
        x <- as.matrix(x)
    else stopifnot(is.atomic(x))
    if (is.data.frame(y))
        y <- as.matrix(y)
    else stopifnot(is.atomic(y))
    .Internal(cov(x, y, na.method, FALSE))
}
<bytecode: 0x0000000002e3b6b8>
<environment: namespace:stats>
```

All this to say: avoid using loops in R (at least those with many iterations and simple tasks). In Chap. 7 we will briefly discuss some ways of speeding up R and in Chap. 4 we will revisit the *apply* functions and show that they are not necessarily really skipping over the loops the way we might think they are. You might also want to read [44] for some additional handy tips. Check the R help for other useful recursive functions such as *lapply, eapply, mapply, rapply, tapply, by...*

Back to our analysis after this *intermezzo*. We have the indexes for SNP and samples. No samples were deemed unusable but 5458 SNP did not pass QC (see Appendix A for details). Let's remove them

```
> genotypes=genotypes[-indexsnp,]
> dim(genotypes)

[1] 49519      83
```

Done! We have already seen all the tools we need for the analysis in the previous chapter. All we have to do is test a model against each individual SNP. For illustration purposes we will use a very simple model, just the trait modeled by the SNP with no other effects. We have no trait data, so let's simply invent some. Our trait is weight

```
> weight=rnorm(83,mean=50,sd=10)
> summary(weight)

   Min. 1st Qu.  Median    Mean 3rd Qu.    Max.
  28.90   43.61   49.81   51.25   59.69   83.41

> sd(weight)

[1] 10.97551

> plot(density(weight),col="blue",
+      main="Density plot of weights")
> abline(v=mean(weight),col="red")
> lines(density(rnorm(83000,mean=50,sd=10)),
+      col="green",lty=2)
```

What did we do? We sampled 83 measures from a normal distribution with a mean of 50 and a (rather large) standard deviation of 10. This is a really nice feature in R—it is very easy to sample from different distributions (e.g., *rnorm, rbinom, rgamma, rbeta, rcauchy, rchisq, rexp, rf...*). Check also *dnorm, pnorm*, and *qnorm* for respectively, density, distribution, and quantiles (for other distributions just replace *d, p*, and *q* with the name of the distribution). The density plot (Fig. 3.8) shows that the data is not perfectly normal due to the small sample size and large SD—at least in comparison to the large sample overlaid on the plot. Does this data make sense? Of course not! There are no effects associated with a particular genotype on any SNP so we would not expect any effect at all. And that's why we are doing this—with almost 50,000 tests we are in multiple testing hell. At a 1% significance level we would expect almost 500 false positives.

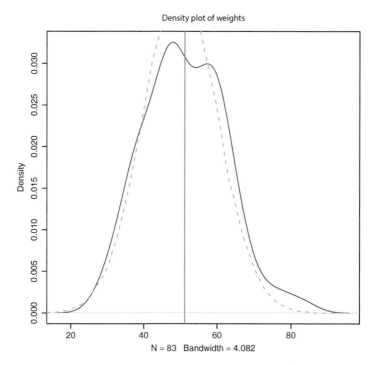

Fig. 3.8 Density plot of 83 animals sampled from a normal distribution with mean 50 and SD 10. The *dotted line* shows the density plot for a sample of 83,000

```
> singlesnp=function(trait,snp)
+ {
+          if (length(levels(snp))>1) lm(trait~snp)
+          else NA
+ }
> results=apply(genotypes,1,
+      function(x) singlesnp(weight,factor(t(x))))
> pvalfunc=function(model)
+ {
+          if(class(model)=="lm") anova(model)[[5]][1]
+          else NA
+ }
> pvals=lapply(results,
+      function(x) pvalfunc(x))
> names(results)=row.names(genotypes)
> pvals=data.frame(snp=row.names(genotypes),
+      pvalue=unlist(pvals))
```

First we create a function to fit weight to the SNP (and return *NA* in case there's nothing to fit). Then we use apply to call the *singlesnp* function for each SNP and store the entire model returned by *lm* in *results* (*results* is of class *list*—an R structure that can essentially hold any type of data). Then we use *lapply* (an *apply* on lists) to extract the *p*-values for the anova test. It's the same as we did before, we are just checking if the SNP explains part of our trait. And then some housekeeping (give names to rows and columns) to make sure we know what is what. Note that this will take a few minutes to run.

How many SNP are significant at a 1 % threshold?

```
> length(which(pvals$pvalue<0.01))
```

```
[1] 552
```

Quite close to the number of false positives we would expect. We will address this issue in the next section. In the meantime we will assume that all results are correct and valid. The next step is to have a look at the effect sizes per genotype of the significant SNP. To avoid working with large tables let's exemplify using only the top five SNP. Start by getting the index for the top 5.

```
> index=sort(pvals$pvalue,index.return=T)[[2]][1:5]
```

Function *sort* with the argument *index.return=T* returns a list of results. First the sorted values, next the sorted indices. That's why we use *[[2]]* (double brackets to index lists) for the indices and then *[1:5]* for the first 5. Since we stored the whole model in *results* all we have to do is extract the coefficients from each of the top SNP and build a table to see what we've got.

```
> estimates=NULL
> for (i in 1:5)
+           estimates=rbind(estimates,
+       coefficients(summary(results[[index[i]]])))
> estimates=cbind(rep(c("AA mean","AB dev","BB dev"),5),
+       estimates,rep(names(results)[index],each=3))
> estimates=data.frame(estimates)
> names(estimates)=c("genotype","effect",
+       "stderror","t-value","p-value","snp")
> for (i in 2:5) estimates[,i]=
+       signif(as.numeric(as.character(estimates[,i])),2)
> print(estimates)
```

```
  genotype effect stderror t-value p-value
1  AA mean  78.00      5.4   14.00 5.1e-24
2   AB dev -24.00      5.8   -4.20 6.5e-05
3   BB dev -30.00      5.6   -5.40 8.0e-07
4  AA mean  53.00      1.5   35.00 2.7e-50
5   AB dev  -5.20      2.2   -2.30 2.1e-02
6   BB dev  18.00      5.2    3.50 8.3e-04
```

```
7     AA mean    46.00      4.4     10.00 1.6e-16
8     AB dev    -0.62       4.8     -0.13 9.0e-01
9     BB dev     9.60       4.6      2.10 4.2e-02
10    AA mean    75.00      9.9      7.60 5.9e-11
11    AB dev   -18.00      10.0     -1.80 8.1e-02
12    BB dev   -27.00      10.0     -2.70 8.8e-03
13    AA mean    48.00      1.9     26.00 1.2e-39
14    AB dev     7.70       2.4      3.20 2.1e-03
15    BB dev    -6.00       3.5     -1.70 9.5e-02
                         snp
1    OAR9_100790876.1
2    OAR9_100790876.1
3    OAR9_100790876.1
4      DU281388_299.1
5      DU281388_299.1
6      DU281388_299.1
7    OAR14_25266721.1
8    OAR14_25266721.1
9    OAR14_25266721.1
10   OAR5_101989167.1
11   OAR5_101989167.1
12   OAR5_101989167.1
13   OAR13_23609874.1
14   OAR13_23609874.1
15   OAR13_23609874.1
```

And that's all there is to it.

So far we have not seen how to use map information in practice. We can exemplify with a common GWAS plot (we will revisit GWAS plots in the next chapter). For each chromosome plot the negative log odds of the p-values for the association test by physical location. We can get the map from our database and merge the p-values with mapping information

```
> map=dbGetQuery(con,
+       paste("select * from snpmap",sep=""))
> merged=merge(pvals,map,by.x=1,by.y=1)
```

Now let's plot the log odds for chromosome 1 (Fig. 3.9). Here we are using the natural logarithm but you would usually use base 10 for log odds (just replace *log* with *log10*).

```
> plot(merged$position[which(merged$chromosome==1)],
+       -log(merged$pvalue[which(merged$chromosome==1)]),
+       xlab="map position",ylab="-log odds",
+       col="blue",pch=20,main="Chromosome 1")
> abline(h=-log(0.01),col="red")
```

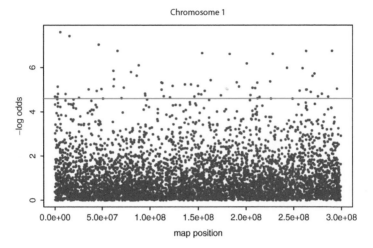

Fig. 3.9 Plot of negative log odds of association test in chromosome 1. The *line* shows the significance threshold for 0.01

Have a look at the package *lattice* for building chromosome plots, as for example those shown in appendix A.

3.6 Multiple Testing

Multiple testing can be quite complicated—there's even a whole book on the subject almost entirely devoted to genomic applications [25]. We will only discuss the very simple Bonferroni correction and some key points. Bonferroni is easy, just divide the desired significance level for the test by the number of tests. In our case we would only accept SNP with a *p*-value under

```
> 0.01/length(pvals[,1])
```

```
[1] 2.019427e-07
```

The problem with Bonferroni is that it is overly conservative. SNP are not independent but rather in linkage with each other, this should be taken into account. We could instead monitor the false discovery rate (FDR), e.g., set it to be 5 or 10 % of the significant SNP. Even though there is a lot of discussion about multiple testing the issue is probably over inflated. The truth is we are working with cutoffs and however simple or complicated a correction method we use there will always be some level of uncertainty (and the accompanying false positives or false negatives). Further, a GWAS is not the end of the story—it is the lead into new insights that have to be further pursued, some false positives in our results are probably better than false negatives, in the first case it's just a dead end trail (and more money, of course!). Depending on the objectives of the study, the best approach may be to simply rank the results and work your way from the top down—remember that whatever the correction used, the ranking will still be the same.

But let's see how many of our significant SNP survive multiple testing correction

```
> length(which(pvals$pvalue<0.01))

[1] 552

> length(which(pvals$pvalue<0.01/length(pvals$pvalue)))

[1] 0

> sort(pvals$pvalue)[1:5]

[1] 1.591244e-06 6.905358e-05 8.181758e-05 8.391490e-05
[5] 9.044598e-05
```

None—well in our case that's good. We really would hope that randomly selected trait values would not survive such a stringent filter.

3.7   What Next

This is really just the tip of the iceberg. From here you might want to impute missing genotypes based on LD information or pedigree structure. Single SNP analysis are a good starting point but you will want to test for association using multiple SNP and/or haplotypes. Also try more robust methods for fitting SNP such as random effects in a mixed effects model or a Bayesian approach. We will briefly visit some of these topics in the next chapter but let's leave most of this for *Advanced Analysis of Genomic Data Using R...*

3.8   Useful R Packages

The packages we discussed in the last chapter are all useful here. Add to those:

- *GenABEL*, a comprehensive suite of functions for GWAS and one of the most widely used
- *fdrtool*, handy functions for false discovery rates
- *multtest*, yes you guessed it—multiple testing
- *qtl* for analyzing QTL projects and *bim* for a Bayesian approach
- *GeneticsPed*, some nice functions for handling pedigrees
- *beadarraySNP* has lots of functions and reporting options for Illumina data
- *haplo.stats* for haplotypes
- *snpMatrix* is a flexible package with many functions for association studies and imputation

There are quite a few more—but this should keep you busy for a while!

Chapter 4
Populations and Genetic Architecture

In this chapter we overview additional uses of SNP markers. We extend the single SNP association analysis from the previous chapter and now fit all SNP simultaneously with SNP best linear unbiased prediction (*snpBLUP*). Genomic prediction using *snpBLUP* and *gBLUP* are then discussed. We then overview how to identify signatures of selection, estimate population parameters, measure linkage disequilibrium and relationships between populations. The final section illustrates a practical application for SNP genotypes: parentage testing. Most of these analyses rely on manipulation of matrices—how to work with matrices in R is also discussed.

4.1 Beyond Genome Wide Association Studies

The primary use of SNP arrays has been association studies: identification of genomic regions associated with a trait of interest; the final objective being to identify the causal variants that lead to the phenotype. In the last chapter we performed SNP by SNP analyses but also mentioned that there are issues with this approach, the main ones being false discoveries due to multiple testing and over estimation of the size of effects since all other SNP are disregarded. With quantitative traits that are highly polygenic single SNP regressions will miss out on a lot of *true* signals due to the stringency of the significance testing. In this chapter we will discuss how to fit all SNP simultaneously using *snpBLUP* and compare with single SNP regression results.

While human studies generally have more of a *QTL vibe* to them, in livestock the main objective has been to use SNP data to make predictions of outcome—either predict the phenotype of the individual itself or estimate its breeding value (additive

Electronic supplementary material The online version of this chapter (doi: 10.1007/ 978-3-319-14475-7_4) contains supplementary material, which is available to authorized users.

C. Gondro, *Primer to Analysis of Genomic Data Using R*, Use R!,
DOI 10.1007/978-3-319-14475-7_4

genetic value). The key difference is that in livestock individual SNP effects are not so relevant, what is relevant is how well the SNP are able to predict a phenotype. A method for genomic prediction that does not even estimate individual SNP effects is *gBLUP*—we will use it to predict phenotypes and see that it is an equivalent model to *snpBLUP*.

But there is much more that can be done with SNP data, it has allowed obtaining much better estimates of population parameters such as heterozygosity, inbreeding, effective population sizes, relationships between populations, identification of signatures of selection, and even very practical applications such as parentage testing. We will discuss some of these later in the chapter.

But first, all these methods are heavily reliant on matrix algebra. We will start this chapter with an overview of how to work with matrices in R.

4.2 Matrix Algebra

Readers are probably familiar with the basics of matrix algebra and we will not need more than a broad understanding of it for what we will do in this chapter. If you are unfamiliar with terms such as *inverse, transpose, identity,* or *determinant* it might be worth having a quick look at a linear algebra book (in fact, the *Wikipedia* entry for *Matrix* suffices). But what we do need to know is that matrix manipulations are mostly just repetitive operations performed on the elements of a matrix. A programmer will intuitively resort to using loops for matrix operations but as we already discussed in Chap. 3 these are rather inefficient in R. Let's sidetrack a little and see why loops are inefficient in R.

4.2.1 Loops and Vectorization

This section starts with a rather obvious comment: R is not *C* or *FORTRAN*. Usual programming practices do not always translate well into R. Keep in mind that R is an interpreted language and not a compiled one. Things will always be slower in R than in, e.g., *C* or *FORTRAN*; at best the speed will be the same. On the positive side, most of R itself is written in *C* or *FORTRAN* and many functions have been implemented in these languages (see details in Chap. 7). This naturally suggests that the fewer times the interpreter has to be called upon to then send a command to run a compiled function, the faster the code will execute. But before looking into this, let's look at something that most languages do not like: memory re-allocation. Since R does not force us to declare variable types nor their size before we use them, we tend to forget that as we grow the contents of a variable we have to allocate memory for it. This is time consuming (and fragments the memory). Typical villains are *c, cbind,* or *rbind,* very handy commands but also very slow. Let's illustrate with a vector of one million random numbers that are added 100 times to an initially empty variable

```
> nums=rnorm(1000000)
> numsMat=NULL
> for (i in 1:100) numsMat=cbind(numsMat,nums)
```

This took 27.9 s to complete. If instead we pre-allocate the size of the matrix and then fill in the data it takes only 2.6 s.

```
> numsMat=matrix(NA,1000000,100)
> for (i in 1:100) numsMat[,i]=nums
```

And even better, is to simply create the whole matrix in a single line in 1.4 s.

```
> numsMat=matrix(rep(nums,100),1000000,100)
```

This is orders of magnitude faster and one of the easiest ways to improve performance in R; and probably also one of the most common pitfalls. The last snippet of code above returns to the initial point of this section, that it is better to use functions compiled in a low level language (as *rep* here) and avoid repeated calls to the R interpreter. To further illustrate let's load a dataset of genotypes into R and calculate the allele frequencies. The file (*genotypes.rds*) is in the folder for this chapter in RDS format (RDS is a binary format that is quick to read in R, details in Chap. 7). The function to read an RDS file into R is *readRDS(fileName)*. The data consists of 2,000 individuals genotyped for 10,000 SNP, genotypes are coded as 0, 1, and 2 with 1 being the heterozygotes; and no missing data—all set for fitting an additive effects model (more on this data later). We could use loops to calculate allele frequencies:

```
> geno = RDS("chapter4/genotype.rds")
> freqA=numeric(10000)
> freqB=numeric(10000)
> for (i in 1:10000)
+ {
+    hold=0
+    for (j in 1:2000)
+        hold=hold+geno[i,j]
+    freqB[i]=hold/4000
+    freqA[i]=1-freqB[i]
+ }
```

This is a rather literal translation from *C* or *FORTRAN* and quite inefficient in R (it took 13.3 s). We can make some good progress by using vectorization. Generally R functions are vectorized—they will perform the same operation on a single value or a vector of values using a lower level language. Of course this still implies looping over the vector but if this is done from within a compiled function, the speed gains can be large. For example, replacing the inner loop with the *sum* function which adds all elements of a vector we come down to 0.6 of a second.

```
> freqA=numeric(10000)
> freqB=numeric(10000)
> for (i in 1:10000)
+ {
+     freqB[i]=sum(geno[i,])/4000
+     freqA[i]=1-freqB[i]
+ }
```

And we can do even better with the compiled function *rowSums* and make use of vectorization in the whole process:

```
> freqB=rowSums(geno)/4000
> freqA=1-freqB
```

Here, as the name suggests, *rowSums* calculates the sum for each row in a compiled function, then each element is divided by the number of alleles (two times the number of individuals), finally we calculate the frequency for the other allele by subtracting the frequencies in *freqB* from 1.0 for all SNP. If the rationale behind dividing by 4,000 is not clear, it is because there are 2,000 individuals and genotypes are coded as 0, 1 and 2—if e.g., all individuals for a SNP were homozygous 2, the sum of all genotypes for that SNP would be 4,000 which divided by two times the number of individuals ($2 \times 2,000 = 4,000$) would give an allele frequency of 1.

Now we are down to 0.06 s and only two lines of code. In Chap. 7 we discuss how to write a similar function in $C++$ and call it from R. Note also the vectorization to calculate the frequency of B; it is simply $1 - freqB$ and we did not need to explicitly loop through every element of *freqB* to take the difference.

In summary, R is slow with loops. With a loop you are each time going back and forth through the interpreter. Try to send everything as much as possible as a vector which can then be iterated through using compiled functions. Recall that most of R is written in *C* or *FORTRAN*, it is just a matter of presenting the data more efficiently to these functions. This book clearly cannot cover all the different functions available in R but as a rule of thumb whenever the code involves double loops it is worthwhile to spend some time investigating if there is a low level language function already written that will perform the operation.

At this point it is worth revisiting the *apply* family of functions that were briefly discussed in Chap. 3. Using these functions it is possible to avoid loops and vectorize the task. Use these functions to iterate through a vector (*sapply*), list (*lapply*), matrix (*apply*) or some other object by *applying* a function on their elements. For example *apply* runs a function over the lines or columns of a matrix. These functions are convenient and make for concise coding (albeit not intuitive until you are familiar with the syntax) but they are not always very useful to improve performance. To some extent it is just *loop hiding*, the speed gains are more related to the function that is being applied rather than the *apply* function itself. To illustrate let's run a single SNP regression on a phenotype (additive model) and store the *p*-values. Just invent some values for a phenotype and then use *lm* for a single SNP regression (as we did in the last chapter).

```
> pheno=rnorm(2000,mean=100,sd=10)

> pvals=numeric(10000)
> for (i in 1:10000)
+   pvals[i]=coef(summary(lm(pheno~geno[i,])))[2,4]
```

This took 31.15 s. The downside with this approach is that we make a call to the *lm* function (which is compiled code), return to R, call the function again, and repeat this 10,000 times. Intuition would suggest that *apply* would do a better job.

```
> pvals=apply(geno,1,
+   function(snp) coef(summary(lm(pheno~snp)))[2,4])
```

This takes even slightly longer (31.31 s). In reality we did not vectorize anything; all we did was call *lm* 10,000 times again. The take home message is that the *apply* functions will not automatically vectorize a task, if the code has reached a stage where the only option is to iterate, it will seldom make any difference in terms of speed if a *for* loop or an *apply* function is used.

A last point is that not everything is vectorizable. Some computations may depend on the result from a previous stage (e.g., Bayesian approaches). In these instances it can be worthwhile to write the function in a low level language instead. If you are going to stick with R, try not to repeat any calculations within loops and if memory is not an issue, sometimes it is slightly more efficient to use suboptimal code (e.g., doubling up entire objects with some change) rather than trying to modify specific elements in an object.

4.3   Matrix Operations in R

This rather long break was to understand what R is doing behind the curtains. Now we will see matrix operations in R but keep in mind that in reality this is just making optimal use of vectorization and reducing communication with the interpreter. Matrix operators are quite intuitive in R; start by centering the matrix of genotypes on zero (i.e., all genotypes minus 1.0).

```
> M=geno-1
```

Now the heterozygotes are zero and the homozygotes are -1 and 1 (this will be useful later). Again, a programmer would think in terms of double loops (e.g., an outer loop for the rows and an inner loop for the columns) to implement this subtraction; R does this automatically for us. Similarly, if two matrices have the same dimensions we can do an element-wise multiplication (i.e., all $a[i, j] * b[i, j]$ multiplications) with e.g.

```
> M2=M*geno
```

And to multiply a matrix by a vector (e.g., the phenotypes—*pheno*)

```
> M2=M*pheno
```

This multiplies every *column* in *M* with the values in *pheno*. If the dimensions do not match, R will return a warning message

```
> M2=pheno[1:13]*M
```

```
Warning message:
In pheno[1:13] * M :
  longer object length is not a multiple
          of shorter object length
```

Pay some attention to the direction: down the *rows* and across the *columns* (easy to swap things around and with square matrices there will be no warning to indicate a problem). Of course, the above works with all numeric operators: plus, minus, division...

Now for some *real* matrix operations. To transpose a matrix

```
> Mt=t(M)
> dim(M)
```

```
[1] 10000   2000
```

```
> dim(Mt)
```

```
[1]   2000 10000
```

And for matrix multiplication

```
> MtM = Mt %*% M
> dim(MtM)
```

```
[1] 2000 2000
```

Recall that the number of columns of the left matrix has to be the same as the number of rows of the right matrix (here 10,000) and the resulting matrix will be the number of rows on the left matrix and the number of columns on the right matrix (here *2,000 × 2,000*). Each entry in the new matrix is the *dot product* of the rows on the left by the columns on the right. The first element in matrix *MtM* is

```
> MtM[1,1]
```

```
[1]  5019
```

which is obtained by

```
> sum(Mt[1,]*M[,1])
```

```
[1]  5019
```

Here we multiplied *SNP* × *SNP*, we could also do *sample* × *sample* by inverting the order of the two matrices.

```
> MtM = M %*% Mt
> dim(MtM)
```

```
[1] 10000 10000
```

You will have noticed that this is rather slow. If there is an extensive need to use matrices, *Revolution R* (an *enhanced* version of R) is much faster. *Revolution R* was developed by Revolution Analytics and is a commercial, enhanced version of R; just recently the program was made freely available, it can be used exactly in the same way as R (with some additional *tweaks*). See details at *revolutionanalytics.com*. In R the first matrix multiplication took 46.2 s and in Revolution R only 0.85 of a second. The remainder of the chapter will be much faster using Revolution R. A faster way to get the same result is with the *crossprod* function which calculates the product of the transpose of a matrix with a second matrix (*crossprod(A,B)*) or the transpose of a matrix with itself (*crossprod(A)*). This is about twice as fast in R, no real difference in Revolution R. We will not use *crossprod* in the chapter for clarity purposes, but remember to use it with real applications.

```
> # same as t(M)%*%M
> MtM = crossprod(M)
```

The last operator we need is *solve*, for matrix inversion. We need this to get a matrix from one side of an equation to the other; e.g., from $y = Ax$ to $x = A^{-1}x$.

```
> Minv=solve(MtM)
```

solve can also be used to get the values of x in $y = Ax$ with $x = solve(A,y)$ which is the same as $x = solve(A) \%*\% y$.

Additional functions useful for working with matrices are: *upper.tri* and *lower.tri* to extract the upper and lower elements of a matrix; *diag* for diagonal elements; *eigen* to calculate eigenvalues and eigenvectors and *svd* for singular value decomposition. We will illustrate the use of these functions throughout the chapter, but also check the help files for additional parameters.

4.4   SNP Best Linear Unbiased Prediction

While single SNP regression analyses are informative and simple to do, they have some drawbacks as previously mentioned. In this section we will use *snpBLUP* to simultaneously estimate all SNP effects.

The data for this section is simulated; there are 10,000 SNP for 2,000 individuals (*gwasData.rds*) with the SNP coded as *0, 1* and *2* with no missing genotypes. The SNP are already ordered by map position (coordinates in *map.txt* file) and there is also a file with phenotypes (*phenotypes.txt*) in the same order as the samples in the genotypes file. The population is completely unrelated, genotypes are in linkage equilibrium (i.e., there is no linkage between SNP), allele frequencies are roughly the same at 0.5; and there are ten *true* QTL in the data which are actual SNP (i.e., there will be no loss of linkage phase between marker and QTL due to recombination). The phenotypes are *mean centred* and all known fixed effects have been accounted for (equivalent to the residuals after adjusting for all confounding factors). It is a somewhat unrealistic dataset but useful for our purposes. Before fitting all SNP simultaneously let's revisit the marker-by-marker analysis and build up from there. Start by reading in the data and fitting a linear model using *lm*.

```
> # read in data - simulated with 10 QTL
> gwas=readRDS("chapter4/gwasData.rds")
> pheno=read.table("chapter4/phenotypes.txt",
+   header=T,sep="\t")$Pheno
> map=read.table("chapter4/map.txt",header=T,sep="\t")
> dim(gwas)

[1] 10000  2000

> # single SNP regression with lm
> effect=numeric(10000) # effect sizes (coefficients)
> pval=numeric(10000) # p-values
> for (i in 1:10000)
+ {
+   res=coef(summary(lm(pheno~gwas[i,])))[2,c(1,4)]
+   effect[i]=res[1]
+   pval[i]=res[2]
+ }

> effect[1:4]

[1] -0.4989296  0.1690589  0.1876288 -0.1043797

> pval[1:4]

[1] 0.01431413 0.40415900 0.35829852 0.60490297
```

This is quite similar to what we did in Chaps. 2 and 3, but in those chapters a *genotype* model was used; i.e., genotypes were fitted as *factors* and we estimated effects of all other genotypes as a deviation in relation to a *baseline* genotype. This has the advantage of not making assumptions about the mode of inheritance (dominant, recessive...). But we can also explicitly model different modes of inheritance. Instead of the analysis of variance method used so far, a simple linear regression approach can be used. A linear regression, in its simplest form, models the relationship between two numeric variables, e.g., how does variable *x* change when variable *y* changes. In GWAS terms, it translates to: *how much does phenotype (y) change when genotype (x) changes*. Genotypes are now coded as numeric values (not factors) according to a predefined mode of inheritance that the researcher wants to test. For example, an additive model (which is what we fitted here) assumes that alleles have a *dose* effect; one allelic variant has no effect (e.g., *A*) on the trait and the other has an effect (e.g., *B*), a heterozygous individual (*AB*) will express the effect and a homozygous individual with two copies of the allele (*BB*) will express twice the value of the effect (hence the term *additive*). This naturally suggests coding SNP genotypes as *0* (*AA*), *1* (*AB*), and *2* (*BB*). The importance of additive models will become clearer later on.

Returning to the analysis with the SNP data, two variables (*effect* and *p-val*) were created to store the values of the coefficients (additive effects) and the *p*-values from the regressions. The *effects* are the estimates of how much a single copy of the *B* allele changes the value of the trait—it is the *allele substitution effect* (not an entirely appropriate use of the term but within our context it is sensible). Note that effects can be positive or negative (*B* allele increases or decreases the phenotypic measurement in relation to *A* allele). Further, these are just relative estimates (one allele in relation to the other) and the sign is only due to the coding used for the genotypes (code *BB* as *0* and *AA* as *2* instead, the signs will invert). This is really just a *line fitting* exercise; let's compare the results with the highest and lowest *p*-values to make this point clearer:

```
> which(pval==min(pval))

[1] 8577

> which(pval==max(pval))

[1] 7296
```

The most and least significant SNP are 8577 and 7296, these can be plotted with

```
> # split screen for two plots
> par(mfrow = c(2,1))

> # most significant SNP
> plot(gwas[8577,],pheno,
+   xlab="genotypes", ylab="phenotypes",
+   main=paste("effect size:",round(effect[8577],2)))
> mod=lm(pheno~gwas[8577,])
> abline(mod,lwd=2,col="blue") # add regression line

> # least significant SNP
> plot(gwas[7296,],pheno,
+   xlab="genotypes",ylab="phenotypes",
+   main=paste("effect size:",round(effect[7296],2)))
> mod=lm(pheno~gwas[7296,])
> abline(mod,lwd=2,col="blue")
```

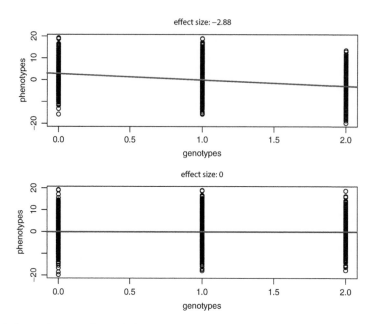

Fig. 4.1 Regression analysis results for the most and least significant SNP. Model fitted is *additive* with genotypes coded as *0, 1,* and *2*. The *line* shows the regression *fit*

Figure 4.1 shows the most and least significant SNP. There is a strong additive association with the first SNP—the regression line has a negative inclination (slope) and an estimated effect size of −2.88; with the second SNP the regression line is flat and the estimate of effect size is zero. Later we will check if the first SNP is one of the *true* QTL. Note that the differences (in the significant SNP) are rather subtle, SNP effects of quantitative traits are seldom visually obvious. We used the *lm* function for the regression analysis and what it did was fit a simple model of the form $y = a + bx$, basically given y (phenotypes) and x (genotypes), it used least squares to estimate the values of a (intersect) and b (slope, SNP effect). For the first SNP this is

```
> y=pheno
> x=gwas[1,]
> mod=lm(y~x)
> summary(mod)

Call:
lm(formula = y ~ x)

Residuals:
     Min        1Q    Median        3Q       Max
-19.2988   -4.5255   -0.0773    4.3939   19.1710
```

```
Coefficients:
            Estimate Std. Error t value Pr(>|t|)
(Intercept)    0.5069     0.2512    2.018   0.0437 *
x             -0.4989     0.2035   -2.451   0.0143 *
---
Signif. codes:   0 *** 0.001 ** 0.01 * 0.05 .

Residual standard error: 6.376 on 1998 degrees of freedom
Multiple R-squared: 0.002999, Adjusted R-squared: 0.0025
F-statistic: 6.01 on 1 and 1998 DF, p-value: 0.01431
```

We used variable y for the phenotypes and x for the genotypes of the first SNP to match with the notation used above. For this SNP the estimates of intercept (a) and slope (b) are respectively *0.5069* and *−0.4989*. The objective of this text is to be applied, so purposefully numerical details have largely been avoided, but it is easier to understand how to do *snpBLUP* if the regression is built up from scratch. The most common way of solving for a and b is with least squares (it minimizes the sum of the squared residuals—values for a and b that give the smallest difference between observed and predicted values of y). The notation for this would be

$$\hat{b} = \frac{\sum_{i=1}^{n}(x_i - \bar{x})(y_i - \bar{y})}{\sum_{i=1}^{n}(x_i - \bar{x})^2} \tag{4.1}$$

$$\hat{a} = \bar{y} - \hat{b}\bar{x} \tag{4.2}$$

The estimate of b is just the sum of the product of the deviations from the mean for x and y, divided by the square of the sum of the deviations from the mean of x. And the estimate of a is the mean of y minus the product of the mean of x with the estimate of b. In R the code is

```
> b=sum(((x-mean(x)) * (y-mean(y)))/sum((x-mean(x))^2)
> a=mean(y)-b*mean(x)
> a

[1] 0.5069125

> b

[1] -0.4989296
```

Note that the results are exactly the same as those from *lm*. The intercept and the slope are the two parameters needed to define the line through the data and can be used to predict values of y given x ($y = a + bx$). The *predict* function in *lm* can be used or simply use the coefficients directly. There are only three genotypes, so there can only be three values for y:

```
> pred=predict(mod)
> pred[1:4]
```

```
            1                  2                  3                  4
 0.007982874   0.506912508  -0.490946760   0.506912508

> yhat=a+b*x
> yhat[1:4]

[1]   0.007982874   0.506912508  -0.490946760   0.506912508
```

Without details and just for completeness (an excellent text for regression in R is [100]), the same R^2, *F-values*, and *p-values* can be obtained directly with

```
> SST = sum((y - mean(y))^2)
> SSR = sum((yhat - mean(y))^2)
> SSE = sum((y-yhat)^2)

> R2=SSR/SST
> Fval=SSR /(SSE/(length(y)-2))
> Pval=1-pf(Fval,1,(length(y)-2))
> c(R2,Fval,Pval)

[1]  0.00299874 6.00950385 0.01431413
```

This is what the loop using *lm* did behind the scenes. The analysis could be replicated with a loop using the *manual* approach we just overviewed or we could instead use matrix algebra (note that this is still single SNP regressions—just doing it with matrices). The equivalent matrix notation is

$$y = X\beta + e \qquad (4.3)$$

where X is a design matrix that assigns the *mean* (a—intercept) and number of SNP alleles (b—slope, effect) to the phenotype records (y). β are the regression coefficients ($\beta_0 = a$ and $\beta_1 = b$) and e is the random error term assumed to be $e_{ij} \sim N(0, \sigma_e^2)$ where σ_e^2 is the variance error. The design matrix X needs two columns (one for each β), the first column is for the intercept, which is the same for all individuals, so it is simply a column of *ones*; the second column is for the *dose* effect (slope) and will be the number of copies of one of the alleles (here, allele B) for each individual. For this example X is a matrix of 2,000 rows by 2 columns, filled with *1's* in the first column and the genotypes (*0, 1* or *2*) in the second column. We want to solve for β which can be estimated with

$$\begin{bmatrix} \hat{\beta}_0 \\ \hat{\beta}_1 \end{bmatrix} = [X'X]^{-1} [X'y] \qquad (4.4)$$

This can be ported to R as

```
> y=pheno
> X=matrix(0,2000,2)
```

```
> X[,1]=1 # fill with ones for intercept
> # store output
> interceptM=numeric(10000) # beta0 or a (intercept)
> effectM=numeric(10000) # beta1 or b (slope)

> # solve for all SNP
> for (i in 1:10000)
+ {
+   X[,2]=gwas[i,]
+   XtX=t(X)%*%X
+   lhs=solve(XtX)
+   rhs=t(X)%*%y
+   sol=lhs%*%rhs
+   interceptM[i]=sol[1,1]
+   effectM[i]=sol[2,1]
+ }
```

The matrix *sol* is a 2×1 matrix with solutions for the βs. The values of $\hat{y}$ (predicted y) for, e.g., *SNP 1* can be obtained with

```
yhat=interceptM[1]+effectM[1]*gwas[1,]
```

The results are identical to those from *lm* or from using least squares to estimate the intercept and slope.

```
> # SNP effect from lm
> effect[1:4]

[1] -0.4989296   0.1690589   0.1876288  -0.1043797

> # SNP effect from matrix
> effectM[1:4]

[1] -0.4989296   0.1690589   0.1876288  -0.1043797
```

Now we are finally ready to fit all SNP simultaneously using *snpBLUP* (ridge regression). This approach uses the same models suggested for genomic prediction [75] and involves fitting SNP as random effects. While *snpBLUP* solves the problem of multiple testing and over estimation of SNP effects, it does strongly shrink back the estimates of SNP effects. Effects are treated as random samples from a normal distribution and assumed to have equal variance (with many SNP the variance has to be very small). Alternative methods use different variance estimates for each individual SNP (e.g., Bayesian methods; see [34] for an overview with R examples). In its simplest form *snpBLUP* is

$$y = \mu + Xg + e \tag{4.5}$$

where μ is the trait mean, X is the design (genotypes) matrix linking SNP effects to phenotypes, and g is the vector of additive genetic effects. Following [52], the mixed model equations to solve for g (SNP effects) are

$$\begin{bmatrix} \hat{\mu} \\ \hat{g} \end{bmatrix} = \begin{bmatrix} 1'_n 1_n & 1'_n X \\ X' 1_n & X'X + I\lambda \end{bmatrix}^{-1} \begin{bmatrix} 1'_n y \\ X'y \end{bmatrix} \tag{4.6}$$

This takes a similar form to what we did before but X now includes all SNP (the matrix of genotypes) and there is the parameter λ which is $\lambda = \sigma_e^2 / \sigma_g^2$ (ratio of residual by genetic variance). The σ_g^2 (SNP specific variance) value is usually unknown and can be approximated as σ_a^2 / m (the additive variance divided by the number of markers). The additive and residual variances can be inferred from the heritability of the trait: $\sigma_a^2 = \sigma_p^2 h^2$ and $\sigma_e^2 = \sigma_p^2 (1 - h^2)$. These concepts will be discussed in the next section, for now all we need is to know that the heritability of the trait for this data is 0.5.

A point with this estimate of λ is that it does not take into account differences in allele frequencies—not an issue in this simulated dataset since frequencies are all the same—but a preferable estimate of λ is

$$\lambda = 2 \sum p_i (1 - p_i) * (\sigma_e^2 / \sigma_a^2) \tag{4.7}$$

where p is a vector of the frequencies of one of the alleles.

The equations for the SNP solutions are not straightforward at first glance. However it becomes clearer once they are broken down into their individual components. First, if the trait is mean centered and adjusted for all fixed (unrealistic but simplifies the concept) effects, there is no need to estimate a trait mean, and the solution reduces to

$$\begin{bmatrix} \hat{g} \end{bmatrix} = \begin{bmatrix} X'X + I\lambda \end{bmatrix}^{-1} \begin{bmatrix} X'y \end{bmatrix} \tag{4.8}$$

where X is the matrix of genotypes adjusted for allele frequencies $(X - 2p)$, λ is a scalar value calculated as above and I is an identity matrix; i.e., calculate $X'X$ and then add the value of λ to the elements in the diagonal of $X'X$. The other term is the transpose of X multiplied by the phenotypes (y). If the trait mean needs to be estimated, the other terms are included. 1_n is a vector of *ones* of length n (number of samples); $1'_n 1_n$ is simply the number of samples, $1'_n X$ and $X' 1_n$ are the transpose of each other (i.e., same values) and link the *mean* with the *genotypes* (these values are generally so small that a vector of zeros instead will give the same results). The last term, $1'_n y$ is simply the sum of y. In the R code below, matrix algebra is used but the components of the matrix can be built in any suitable way.

```
> # snpBLUP
> h2=0.5 # heritability
> y=pheno

> p=rowMeans(gwas)/2 # frequency of second allele
```

```
> # lambda (equation 4.7)
> d=2*sum(p*(1-p))
> ve=var(y)*(1-h2) # residual variance
> va=var(y)*h2 # additive variance
> lambda = d * (ve/va)
> # equivalent for lambda:
> #lambda=(1-h2)/(h2/d)

> # snpBLUP (equation 4.6)
> X=t(gwas-(p*2)) # freq. adjusted X matrix
> XtX=t(X) %*% X  # X'X
> diag(XtX)=diag(XtX)+lambda # X'X+lambda

> ones=rep(1,length(y)) # 1n
> oto=t(ones)%*%ones  # 1n'1n
> otX=t(ones)%*%X # 1n'X
> Xto=t(X)%*%ones # X'1n
> # build full matrix:
> lhs=rbind(cbind(oto,otX),cbind(Xto,XtX))

> oty=t(ones)%*%y # 1n'y
> Xty=t(X)%*%y # X'y
> rhs=rbind(oty,Xty)
> effectBLUP=solve(lhs)%*%rhs # SNP solutions
```

Here *rbind* and *cbind* were used to build the matrices—this is inefficient as previously discussed, better to pre-allocate the matrix and then fill it in. *snpBLUP* is computationally intensive and very demanding on memory, large datasets will be unfeasible on commodity machines. This data was already mean centered, so the estimate for trait mean is zero (below). The mean is the first value in the solutions matrix (*effectBLUP*). Rerun the example with, e.g., $y = pheno + 23.7$ to test.

```
> head(effectBLUP)
```

```
             [,1]
[1,]  -3.259863e-16
[2,]  -4.892722e-02
[3,]   1.853745e-02
[4,]   1.145339e-02
[5,]  -7.280290e-03
[6,]  -1.749794e-02
```

Without fitting the mean, the same SNP solutions can be obtained with

```
> # without fitting mean (equation 4.8)
> XtX=t(X) %*% X
```

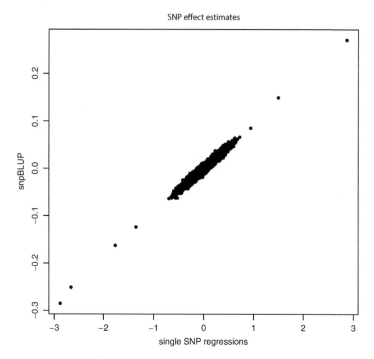

Fig. 4.2 Estimates of SNP effects from single SNP regressions and *snpBLUP*. The correlations are high but there is strong shrinkage of effect sizes with *snpBLUP*

```
> diag(XtX)=diag(XtX)+lambda
> Xty=t(X)%*%pheno
> effectBLUP2=solve(XtX)%*%Xty # SNP solutions
```

Conditional on other SNP, *snpBLUP* shrinks the estimates of effects. The extent of this shrinkage can be seen by comparing the effects from the single marker regressions with *snpBLUP* (Fig. 4.2).

```
> # just to make variable names more meaningful
> effSingle=effect
> effBlup=effectBLUP2

> plot(effSingle,effBlup,
+   xlab="single SNP regressions",
+   ylab="snpBLUP",
+   main="SNP effect estimates",pch=20)

> cor(effSingle,effBlup)

            [,1]
[1,]  0.9779384
```

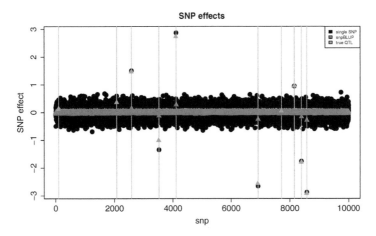

Fig. 4.3 True QTL effects and estimates of SNP effects from single SNP regressions and *snpBLUP*

To finish this section we can compare the results from both methods with the *truth*. 10 QTL were simulated in the data (the size of effects and which SNP can be found in the *trueQTL.txt* file). A plot of effects is informative (Fig. 4.3).

```
> # file with true QTL effects
> QTL=read.table("chapter4/trueQTL.txt",
+   header=T,sep="\t")
> head(QTL)

     QTLval indexQTL
1 -1.0041147     3524
2  0.1370017      100
3  0.9496635     8152
4 -1.7958145     8392
5  0.3614149     2077
6  2.7391442     4112

> plot(effSingle,pch=20,cex=2.5,xlab="snp",
+   ylab="SNP effect",main="SNP effects")
> abline(v=QTL$indexQTL,col="gray")
> points(effBlup,col="red",pch=17,cex=1.5)
> points(QTL$indexQTL,QTL$QTLval,col="green",
+   pch=17,cex=1.5)
> legend("topright",c("single SNP","snpBLUP","true QTL"),
+   fil=c("black","red","green"),cex=0.8)
```

Figure 4.3 reiterates the strong shrinkage of *snpBLUP*. The single marker regression did a better job at approximating the *true* effects in the simulated data but notice how three of the *true* QTL effects are overestimated and many other non-associated SNP show sizeable effects. *snpBLUP* underestimated effects but the non-associated SNP are closer to zero. Methods are reasonably comparable; both identified 6 correct QTL and missed 4 (see below). The trait is not very polygenic with only ten QTL—not an ideal scenario for *snpBLUP*. Association significance can be tested in the same way as in Chap. 3 and can then be compared with the *true* known QTL.

```
> # significant SNP after FDR correction
> # single regression analysis
> sigSNP=which(pval<0.01/length(pval))
> length(sigSNP)

[1] 6

> # number of true QTL found
> length(intersect(sigSNP,QTL$indexQTL))

[1] 6

> # just an approximation based on a t-distribution
> pvalBlup=2*pt(-abs(effBlup/sd(effBlup)),
+ df=length(effBlup)-1)
> sigSNP=which(pval<0.01/length(pval))
> length(sigSNP)

[1] 6

> # number of true QTL found
> length(intersect(sigSNP,QTL$indexQTL))

[1] 6
```

Both methods found the same six true QTL and missed four of them. Before discussing genomic prediction in the next section, a useful and common plot in association studies is the *manhattan plot* (Fig. 4.4). It is the same plot as we did in Chap. 3 for the GWAS results but there only one of the chromosomes was used. The R package *qqman* has a function to create manhattan plots, but it not too hard to build manually:

```
> # manhattan plot
> library(made4) # for colours
> mbcol=as.factor(map$chrom)
> chroms=unique(map$chrom)
> mapcol=getcol(length(chroms))
> chromseps=numeric(length(chroms))
> xdist=length(map[,3])
> cum=0
> chrpos=numeric(length(chroms))
> for (i in 1:length(chroms))
```

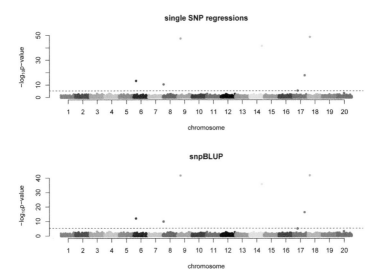

Fig. 4.4 Manhattan plots for single SNP regressions and *snpBLUP*

```
+ {
+       index=which(map[,2]==chroms[i])
+       chromseps[i]=index[length(index)]
+       xdist[index]=map[,3][index]+cum
+       cum=cum+map[,3][index[length(index)]]
+       chrpos[i]=cum
+ }

> chrpos[2:length(chroms)]=
+   chrpos[2:length(chroms)]-
+   ((chrpos[2:length(chroms)]
+   -chrpos[1:(length(chroms)-1)])/2)
>   chrpos[1]=chrpos[1]/2

> par(mfrow = c(2,1))

> # single SNP regressions
> plot(xdist,-log10(pval),col=(mapcol[mbcol]),pch=20,
+   xlab="chromosome",
+   ylab=expression(paste("-",log[10],"p-value",sep="")),
+   axes=F,main="single SNP regressions")
> abline(h=-log10(0.05/nrow(map)), lty=2)
> axis(1, at=chrpos, labels=chroms,las=1)
> axis(2,)
```

```
> # snpBLUP
> plot(xdist,-log10(pvalBlup),col=(mapcol[mbcol]),pch=20,
+   xlab="chromosome",
+   ylab=expression(paste("-",log[10],"p-value",sep="")),
+   axes=F,main="snpBLUP")
> abline(h=-log10(0.05/nrow(map)), lty=2)
> axis(1, at=chrpos, labels=chroms,las=1)
> axis(2,)
```

The packages *biRR*, *BLR*, and *rrBLUP* can be used for ridge regression and *snpBLUP*. We can get almost the same results with the function *mixed.solve* in *rrBLUP*.

```
> library(rrBLUP)
> sol=mixed.solve(y,X)
> effPack=sol$u
> cor(effPack,effBlup)

          [,1]
[1,]  0.9984692
```

Much easier, but not very informative about the underlying methodology. Keep the SNP effect estimates from the three methods (the variables *effSingle*, *effBlup*, and *effPack*), we will need them for the next section.

```
> out=data.frame(single=effSingle,snpblup=effBlup,
+   rrblup=effPack)
> write.table(out,"chapter4/effects.txt",
+   quote=F,sep="\t",row.names=F)
```

4.5   Genomic Prediction

The objective of genomic prediction is to make use of genotypic data to predict phenotypic outcomes, for instance disease susceptibility or prognostics in humans; or production traits in livestock. *snpBLUP* from the previous section can be used for this or the equivalent model *gBLUP*. But first, we saw in the previous section that not all of the *true* QTL were identified as associated with the trait (six out of ten). Looking more closely at the effect sizes of the QTL we see that the six largest ones were identified (coincidentally those with an effect larger than 1) but all QTL with effects below 1 were not. This of course relates to the power of the experiment: with more data, more subtle effects can be identified. However most quantitative traits are polygenic with some QTL of large(ish) effect but many of small effect; the latter cannot be identified in association studies because they do not reach the significance threshold (studies are never large enough to pick these out). These traits are commonly referred to as *complex traits*, i.e., they do not exhibit a simple Mendelian inheritance pattern. A Mendelian trait has either a single or a small number of genes controlling the phenotypic expression, while polygenic traits are controlled by multiple genes (sometimes in very large numbers). Polygenic inheritance, gene–gene interactions and gene–environment interactions contribute to the complexity of a quantitative trait: an observed phenotype is the final expression of all these elements and their interactions with each other. Many important traits in biology, medicine, and agriculture are *complex* in the sense that they exhibit continuous variation and complicated genetic inheritance patterns.

Part of the variation observed in quantitative traits is due to the underlying genetics an individual inherited from its parents, and the other part is due to environmental influences. In other words, both genes and environment contribute to the observed variation in traits or phenotypes. This variation can be partitioned into genetic and non-genetic components (this partitioning is done with statistical methods such as an analysis of variance); thus the observed phenotypic variance can be expressed as the sum of the unobserved genetic variance and the environmental variance:

$$\sigma_P^2 = \sigma_G^2 + \sigma_E^2 \tag{4.9}$$

Here, σ_P^2 is the phenotypic variance while σ_G^2 and σ_E^2 are, respectively, the genetic and environmental variances. The genetic variance can in turn be partitioned into contributions from additive genetic effects (individual contribution of alleles—σ_A^2), dominance effects (interaction between alleles at the same locus—σ_D^2), and epistasis (interaction between alleles at different loci—σ_I^2):

$$\sigma_G^2 = \sigma_A^2 + \sigma_D^2 + \sigma_I^2 \tag{4.10}$$

Heritability, arguably the most common genetic parameter, is defined as the ratio of these variances. It captures the proportion that genes and environment contribute to the variation of a particular trait, and was first introduced by Sewall Wright and Ronald Fisher in the early twentieth century. Broad sense heritability, denoted by H^2, is the proportion of total phenotypic variation due to all genetic effects. It is defined as:

$$H^2 = \sigma_G^2 / \sigma_P^2 \tag{4.11}$$

On the other hand, narrow sense heritability, denoted by h^2, is the proportion of the total phenotypic variation that is attributable only to the additive genetic effects. A parent can pass only one copy of its genes to its offspring; dominance effects cannot be passed to the offspring—it would require sharing both chromosomes and their respective alleles for dominance effects to be heritable; epistasis decays from one generation to the next due to recombination. Narrow sense heritability is defined as:

$$h^2 = \sigma_A^2 / \sigma_P^2 \tag{4.12}$$

To dissect a complex trait we first need to know what proportion of the total phenotypic variance is determined by genetics; and then we can try to find the underlying genetic loci that controls this variation. This is the rationale behind the use of λ in *snpBLUP*. Note that with h^2 the phenotypic variance is split into the additive variance and the *rest*; the latter is not only environmental variance but also all the other non-additive genetic effects (difference of $\sigma_G^2 - \sigma_A^2$); in practice additive models try to explain additive variance and narrow sense heritability (σ_A^2, h^2); they do not capture the full extent of genetic effects on a trait nor the broad sense heritability (σ_G^2, H^2). Herein we will focus only on additive models for prediction purposes and using the simplification: $\sigma_P^2 = \sigma_A^2 + \sigma_E^2$. This was a short (and rather simplified) overview of a few key concepts in quantitative genetics, readers interested in the topic are referred to the classic text of Falconer and Mackay [30].

4.5.1   *Prediction with* **snpBLUP**

As mentioned at the beginning of this section, *snpBLUP* can be used for prediction purposes. Let's do this using the estimates of allele effects calculated from the single SNP regressions and *snpBLUP* from the previous section. Start by reading in the data files: *gwasData.rds* (genotypes file, same as used in previous section), *phenotypes.txt* (trait values for individuals), *effects.txt* (SNP effects calculated in previous section) and *trueGeneticValue.txt* (*true* additive genetic effects of each individual with environmental variance removed).

```
> gwas=readRDS("chapter4/gwasData.rds")
> pheno=read.table("chapter4/phenotypes.txt",
+   header=T,sep="\t")$Pheno
> effect=read.table("chapter4/effects.txt",
+   header=T,sep="\t")
> tgv=read.table("chapter4/trueGeneticValue.txt",
+   header=T,sep="\t")$TGV
```

Given that SNP effects are *known* (and additive), it is simply a matter of multiplying effects by genotypes and summing up across individuals. But first genotypes need to be *scaled* based on allele frequencies as we did before with *snpBLUP*.

```
> p=rowMeans(gwas)/2
> X=t(gwas-(p*2))
```

Phenotypes of individuals can be predicted with

```
> # using effects from snpBLUP
> pred=X%*%effect$snpblup
> head(pred)

            [,1]
[1,]   3.1102863
[2,]   2.9767113
[3,]   1.0998214
[4,]   0.9520851
[5,]   1.7598395
[6,]  -0.9082915
```

To evaluate how well this worked, initially the predicted values can be compared directly with the observed phenotypes.

```
> plot(pheno,pred,ylab="predicted values",
+ xlab="observed phenotypes",pch=20)
> mod=lm(pred~pheno)
> abline(mod,lwd=2,col="blue")

> summary(mod)

Call:
lm(formula = pred ~ pheno)

Residuals:
     Min       1Q    Median        3Q       Max
-2.25237  -0.45804  -0.00115   0.48567   2.61141

Coefficients:
              Estimate Std. Error t value Pr(>|t|)
(Intercept) 9.733e-16  1.603e-02     0.0        1
pheno       4.933e-01  2.511e-03   196.5   <2e-16 ***
---
Signif. codes:  0 *** 0.001 ** 0.01 * 0.05 .

Residual standard error: 0.7168 on 1998 degrees of freedom
Multiple R-squared:  0.9508, Adjusted R-squared:  0.9508
F-statistic: 3.859e+04 on 1 and 1998 DF, p-value: < 2.2e-16

> cor(pheno,pred)

          [,1]
[1,] 0.9750789
```

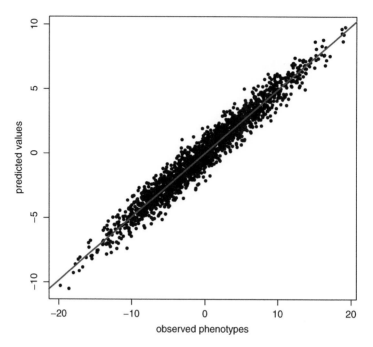

Fig. 4.5 Observed phenotypes versus predicted values using *snpBLUP*

The correlation (square root of the R^2) is very high (0.975) and the fit is very good as shown in Fig. 4.5. Note that the scale of the predicted values is about half that of the observed phenotypes; recall that the heritability of the trait (h^2) is 0.5 (that was set with λ in *snpBLUP*), the SNP effects account for around 50 % of the variation, and the rest is environmental (i.e., not genetic).

This is fine, but it is rather circular. SNP effects were estimated using these same genotypes and phenotypes. Since there is a large number of SNP and not so many records it is relatively easy to find a good fit to the data (the model is over parameterized), but this may not reflect the true genetic values—the unobserved additive genetic values—which is what we really want to know. Since this is simulated data, we know the true genetic value of the individuals (variable *tgv*). Using the true genetic values, we see that our predictions are not so perfect anymore, but still quite acceptable (correlation of 0.662). Notice, in Fig. 4.6, that there is much more scatter in the plot (but the scale is now correct). The phenotypes are the sum of (additive) genetic and environmental effects ($\sigma_P^2 = \sigma_A^2 + \sigma_E^2$); ideally only the true genetic values would be used to estimate SNP effects but they are generally unknown—to increase reliability of SNP effects larger sample sizes are needed to be able to better untangle σ_A^2 from σ_E^2.

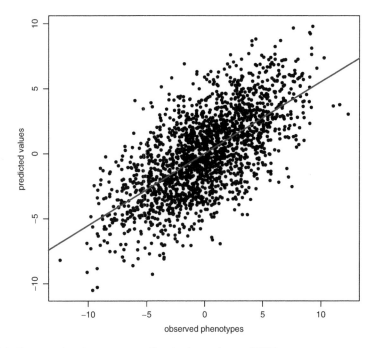

Fig. 4.6 True genetic values versus predicted values using *snpBLUP*

```
> # accuracy with true genetic values (TGV)
> plot(tgv,pred,ylab="predicted values",
+ xlab="observed phenotypes",pch=20)
> mod=lm(pred~tgv)
> abline(mod,lwd=2,col="blue")
> summary(mod)

Call:
lm(formula = pred ~ tgv)

Residuals:
   Min     1Q Median     3Q    Max
-7.677 -1.639 -0.061  1.551  7.870

Coefficients:
             Estimate Std. Error t value Pr(>|t|)
(Intercept) 9.172e-16  5.418e-02    0.00        1
tgv         5.529e-01  1.402e-02   39.43   <2e-16 ***
---
Signif. codes:  0 *** 0.001 ** 0.01 * 0.05 .

Residual standard error: 2.423 on 1998 degrees of freedom
```

```
Multiple R-squared: 0.4376, Adjusted R-squared:  0.4373
F-statistic:  1555 on 1 and 1998 DF, p-value: < 2.2e-16

> cor(tgv,pred)

            [,1]
[1,]  0.6615204
```

Up to this point, the same data was used to estimate effects and evaluate how well these estimates can predict genetic values. In practice, effects should be estimated on one dataset (*discovery*) and evaluated on an independent one (*validation*). For this there are three files in the chapter's folder (*validGeno.rds*, *validPheno.txt* and *validTGV.txt*); let's repeat the prediction steps with this new validation dataset and see how accurate the predictions of genetic values are.

```
> # read in validation data
> validG=readRDS("chapter4/validGeno.rds")
> validP=read.table("chapter4/validPheno.txt",
+   header=T,sep="\t")$Pheno
> validT=read.table("chapter4/validTGV.txt",
+   header=T,sep="\t")$TGV

> validX=t(validG-(p*2)) # X matrix
> validPred=validX%*%effect$snpblup

> # accuracy with true genetic values (TGV)
> plot(validT,validPred,ylab="predicted values",
+   xlab="observed phenotypes",pch=20)
> mod=lm(validPred~validT)
> abline(mod,lwd=2,col="blue")
> summary(mod)

Call:
lm(formula = validPred ~ validT)

Residuals:
    Min      1Q  Median      3Q     Max
-4.8286 -0.8362  0.0157  0.8901  5.2968

Coefficients:
             Estimate Std. Error t value Pr(>|t|)
(Intercept) -0.048810   0.030264  -1.613    0.107
validT       0.099682   0.007892  12.630   <2e-16 ***
---
Signif. codes:  0 *** 0.001 ** 0.01 * 0.05 .

Residual standard error: 1.353 on 1998 degrees of freedom
Multiple R-squared:  0.07394,  Adjusted R-squared: 0.07347
F-statistic: 159.5 on 1 and 1998 DF,  p-value: < 2.2e-16
```

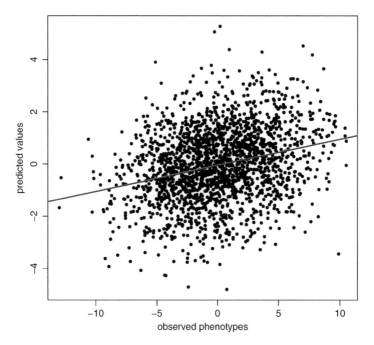

Fig. 4.7 True genetic values versus predicted values using estimated SNP effects in a validation population

```
> cor(validT, validPred)

           [,1]
[1,]  0.2719149
```

There still is some reasonable accuracy of prediction (correlation of 0.272) but not nearly as high as with the discovery population. Figure 4.7 shows a wide scatter and the regression line is much flatter now, note also the scaling of the predicted values—heavily shrunk toward the mean. With a larger *discovery* dataset the estimates of effects would be more accurate and in turn, so would predictions.

Single SNP regressions, as previously mentioned, overestimate effects—particularly of the QTL. Use the SNP effects from the single SNP regressions to make predictions and note how the range of predictions is almost 12× larger than with using the *snpBLUP* effects.

```
> # using effects from single SNP regressions
> predS=X%*%effect$single
> plot(tgv,predS,ylab="predicted values",
+   xlab="observed phenotypes",pch=20)
> mod=lm(predS~tgv)
> abline(mod,lwd=2,col="blue")
```

With this data, the *true* QTL SNP effects are overestimated around 40 % on average, and *snpBLUP* effects are underestimated to just 12 % of the size of the true effects (use the *trueQTL.txt* file to compare QTL effects with the estimates for those SNP).

4.5.2   Prediction with gBLUP

An equivalent model for prediction is *gBLUP*. *gBLUP* is similar to the traditional methodology employed by livestock breeding programs to estimate *breeding values*, but instead of using pedigree information to define the relationships between individuals (a *numerator relationship matrix*) it uses information from a large number of markers. The marker data is used to define *genomic relationships* in a *genomic relationship matrix (GRM)*. The concept is that, if a trait is highly polygenic and controlled by many genes, then the average similarities (and differences) of genotypes between individuals can be used to estimate the sum of their additive effects. So, instead of individually estimating SNP effects and using them to predict the additive genetic value of an individual as we did above, with *gBLUP* the sum of the effects are calculated directly based on the relationships (essentially a covariance matrix between individuals). The model for *gBLUP* is somewhat similar to *snpBLUP*:

$$y = \mu + Zg + e \tag{4.13}$$

with μ being the trait mean, Z being the design matrix allocating records to genetic values, g the vector of additive effects for an individual, and e the vector of random normal deviates. Note that now g refers to individuals and not SNP. Our data is mean centered so, for simplicity, it can be dropped (to fit a mean the exact same approach used in *snpBLUP* can be used). To solve for g (without a mean):

$$\left[\hat{g}\right] = \left[Z'Z + \lambda G^{-1}\right]^{-1} \left[Z'y\right] \tag{4.14}$$

here λ is σ_E^2/σ_A^2. Z is the incidence matrix and y the phenotypes; all that is missing is the genomic relationship matrix (G).

4.5.2.1   Genomic Relationship Matrix

The *GRM* is a relationship matrix with estimates of the similarity among a group of individuals based on SNP data. Apart from its use for phenotypic prediction, it can be used to, e.g., manage inbreeding in livestock or to understand the structure of relationships and genetic distances between and within populations (this will be discussed later in the chapter). The *GRM* is a measure of the additive relatedness between individuals and, in simple terms, can be thought of as a correlation matrix between individuals corrected for variation in allele frequencies. There are various forms of calculating the *GRM*, here the method proposed by van Raden is used [113].

```
> freqAvg=rowMeans(gwas,na.rm=T)
> p=freqAvg/2
> M=gwas-1 # recode matrix as -1, 0, 1
> P=2*(p-0.5) # deviation from 0.5
> W=M-P
> WtW = t(W) %*% W
> d=2*sum(p*(1-p))
> G=WtW/d
> rm(P,WtW,p,freqAvg)
```

First the allelic frequency is calculated (variable p), then the genotypes are scaled to -1, 0, and 1 in matrix M. M is then subtracted from twice the deviation from 0.5 of the allele frequencies $W = M - 2*(p-0.5)$. This corrects for allele frequency differences. G is finally calculated as

$$G = W'W/2\sum_{i}^{n} p_i(1 - p_i) \tag{4.15}$$

The divisor is a scalar that scales G to the allele frequencies of a *base population* (here p was calculated from the data itself).

4.5.2.2   Genomic Prediction with *gBLUP*

Once the *GRM* has been calculated, phenotypes can be predicted with *gBLUP*. The R code is

```
> # lambda
> h2=0.5 # heritability
> y=pheno
> ve=var(y)*(1-h2) # residual variance
> va=var(y)*h2 # additive variance
> lambda = ve/va

> # gBLUP
> Z=matrix(0,ncol(gwas),ncol(gwas))
> diag(Z)=1
> ZtZ=t(Z)%*%Z
> Gil=solve(G)*lambda
> lhs=solve(ZtZ+Gil)
> rhs=t(Z)%*%y
> sol=lhs%*%rhs
```

```
# correlation of snpBLUP X gBLUP predictions
> cor(pred,sol)

         [,1]
[1,] 0.9999999
```

Solutions from *snpBLUP* and *gBLUP* are equivalent. *gBLUP* is more relevant for genomic prediction applications in livestock, but it is equivalent to *snpBLUP* and more computationally efficient since there usually are more markers than samples. SNP effects are not calculated by *gBLUP* but can be obtained by backsolving the *solutions* of individuals. In our case it is simply

```
> backSolve=1/d*W%*%solve(G)%*%sol
```

Predicted values for individuals with *unknown* phenotypes (using the *validation* dataset without the phenotypes) can be obtained by rebuilding the *GRM* with all individuals and then using the relationships between individuals with and without phenotypes to estimate those without.

```
> All=cbind(gwas,validG)
> freqAvg=rowMeans(gwas,na.rm=T)
> p=freqAvg/2
> M=All-1 # recode matrix as -1, 0, 1
> P=2*(p-0.5) # deviation from 0.5
> W=M-P
> WtW = t(W) %*% W
> d=2*sum(p*(1-p))
> Gall=WtW/d
> rm(P,W,WtW,d,p,freqAvg)

> # selection index approach
> index of individuals with no phenotype
> missindex=2001:4000

> # get only samples with phenotypes
> Ginv=Gall[-missindex,-missindex]
> diag(Ginv)=diag(Ginv)+lambda  # add lambda to diagonal
> Ginv=solve(Ginv) # invert
> solAll=Gall[,-missindex]%*%Ginv%*%pheno

> cor(validPred,solAll[2001:4000])

       [,1]
[1,]      1
```

The *pedigree* package has a *gblup* function to predict genetic values and can also calculate the *GRM* (*calcG* function). In this example there were no fixed effects—these would normally be included in the model. The prediction approaches discussed assume that the markers capture all the additive genetic variance and this is only the case if the linkage disequilibrium between markers and QTL is perfect. Note also that large QTL effects are heavily shrunk down and all SNP have a nonzero effect (can be very small though). Bayesian methods will be more appropriate when the trait has QTL of large effects. An in-depth overview of methods for genomic prediction with many R examples is given in [46]. Readers may also find Chap. 11 of [77] useful for a more detailed overview of genomic prediction (oriented toward livestock genomic selection though).

4.6 Population Genetics

Population genetics deals with heredity in populations and the dynamics of the various forces that result in genetic changes. The field revolves around estimation of allele frequencies and how they change over time as populations respond to evolutionary processes such as selection, genetic drift, mutation, and migration. Population genetics also investigates genetic processes such as recombination, linkage and population stratification, as well as environmental adaptation, speciation, and evolutionary relationships. While previously a largely theoretical discipline, the advances in modern molecular technologies enabled population genetics to become a more applied subject since we now have a handle on the structure and variability in populations at the DNA level. While most of the population genetics principles were initially derived from a theoretical framework, we now can effectively test these theoretical models on real experimental data and, due to its solid theoretical foundations, it has become an important toolkit for genomic analysis. Population genetics is important for conservation and ecology studies, it provides insights into evolution and nature but it is also informative in association studies and genomic prediction (e.g., account for population stratification in a GWAS). Here we will focus on the basic population metrics (selection, diversity, linkage, relationships) and some applications using SNP array data, without delving into theoretical details (a good introductory text for population genetics is [51]).

A couple of notes on using SNP array data for population genetics. First the obvious, you can only do this if there is an array available for the species you are interested in working with. Sequence data is becoming quite cheap however and this will not be an issue anymore. Second, arrays are purposely designed to have high minor allele frequencies, not ideal for overall estimates of diversity but adequate in comparative studies. Third, they are subject to *ascertainment bias*, i.e., the data source from which the SNP were selected from can affect results. For example, an array may have been based on sequence data from, e.g., two breeds and it will strongly reflect high levels of diversity in these breeds (because the common SNP were selected for the array), whilst in other breeds these SNP may be less common which would suggest less diversity (the representation of the panel is unbalanced). These last two points are quite relevant and can lead to distorted interpretations if not taken into account.

4.6.1 Signatures of Selection

Selection has shaped all living organisms, simplistically it exposes an advantageous (or disadvantageous) variant and favors its increase (or decrease) in frequency in a population due to differences in reproductive efficiencies. Selection can be due to natural factors as a response to environmental conditions or artificial, such as what occurs in livestock, where human intervention determines which traits are more favorable and individuals carrying better variants for these traits are overrepresented in the matings of the next generation (e.g., the best individual from the gBLUP above is used more often as a breeder).

Positive selection for a particular genetic variant at a locus leads to an increase of its prevalence in a population and in the process leaves unique genetic patterns or signatures in the DNA sequence. Both natural and artificial selection can lead to genomic changes and these are called signatures of selection (*SOS*). SOS studies can reveal regions that were differentially selected in populations of the same species; it is to some extent similar to a GWAS in its objectives: identification of genomic regions associated with a trait. A common approach to identify SOS in the genome of a population is based on quantifying allelic frequency differences between populations using Wright's F_{ST} [54], Tajima's D [108], or Fay and Wu's H [33] statistics. The *pegas* [83] package can be used to calculate F_{ST} as per [117] and D from [108]; *adegenet* has the *pairwise.fst* from Nei [80] and *fstat* is a wrapper for the *hierfstat* package (the latter can be used for hierarchical F_{ST}); see also the *PopGenome* and *snpStats* packages. Other methods to estimate SOS are based on detection of high-frequency long haplotypes with extended linkage disequilibrium; e.g., extended haplotype homozygosity (*EHH*), integrated haplotype score (*iHS*), and standardized ratio of iES—integrated site specific EHH—between two populations (*Rsb*). In R, the *rehh* [38] package can be used to compute these metrics. A good overview of methods for SOS is given in [95].

F_{ST} is the most commonly used method and it is directly related to the variance in allele frequency between populations. The term F_{ST} dates back to 1951 with Sewall Wright [119] but currently there are quite a few derived statistics, all called F_{ST} with the most commonly used being Weir and Cockerham's [117]. F_{ST} methods are based on variance estimates, likelihood maximization, or Bayesian inference. Here we will simply use an analysis of variance approach to test for differences in mean values between groups as proposed by Nicholson et al. [81] which is highly correlated to the F_{ST} of Weir and Cockerham [117]. Parts of this section are adapted from [89], with permission from the publisher.

With SOS studies it is important to identify populations that are relevant for the trait of interest (e.g., populations that have horns versus populations that do not have horns); genetic distances are also important to consider. In the same manner as for GWAS or genomic prediction, the larger the sample size, the better. For SOS it is important to obtain reliable estimates of allele frequencies—ensure populations are balanced (e.g., in livestock no over representation of some sires) and representative of the diversity in the populations. To illustrate we will use SNP data (*sosData.rds*) from 3 different cattle breeds (hanwoo, angus, and brahman) with 25 animals genotyped per breed. The genotypes are part of a single chromosome (around 8,000 SNP) coded as 0, 1, and 2 with no missing values. Note that the genotypes are for illustration purposes only; they are a small subset from a high density panel and in high linkage disequilibrium which is not ideal for these kinds of study. The data is already organized so that the SNP are ordered based on their physical location on the chromosome (rows) and also ordered by breed (columns).

The first thing we need to do is calculate the allele frequencies for each breed.

```
> sos=readRDS("chapter4/sosData.rds")
> dim(sos)

[1] 7763    75

> # matrix to store freqs of each SNP in each pop
> M=matrix(NA,nrow(sos),3) # 3 populations
> colnames(M)=c("hanwoo","angus","brahman")

> # hanwoo
> M[,1]=apply (sos[,which(colnames(sos)=="hanwoo")],
+    1,function(x)  sum(x)/(length(x)*2))

> # angus
> M[,2]=apply (sos[,which(colnames(sos)=="angus")],
+    1,function(x)  sum(x)/(length(x)*2))

> # brahman
> M[,3]=apply (sos[,which(colnames(sos)=="brahman")],
+    1,function(x)  sum(x)/(length(x)*2))

> head(M)

      hanwoo angus brahman
[1,]    0.72  0.02    0.24
[2,]    0.42  0.94    0.00
[3,]    0.58  0.94    1.00
[4,]    0.42  0.36    0.10
[5,]    0.42  0.82    0.92
[6,]    0.42  0.14    0.06
```

F_{ST} is quite simple to calculate in R (or at least this version) and, for each SNP in a population, it is simply the squared deviation (remember it is a variance, hence squared) of the average frequency in that population from the average frequency across all populations divided by the allelic frequency variance ($p*q$). In code this is:

```
> # average allele frequency across populations
> meansB=rowMeans(M)
> alleleVar=meansB*(1-meansB) # p*q variance
```

```
> # deviation of each population from mean
> meanDevB=M-meansB
> # deviation squared divided by var
> FST=meanDevB^2/alleleVar
> head(FST)
```

```
              hanwoo        angus      brahman
[1,]  0.703374419  0.42756112  0.03414831
[2,]  0.004483501  0.95570301  0.82926829
[3,]  0.502976190  0.07440476  0.19047619
[4,]  0.077401372  0.02144082  0.18031732
[5,]  0.446428571  0.04960317  0.19841270
[6,]  0.277582001  0.02710762  0.13120087
```

We can now plot the results for each breed, e.g., hanwoo and try to identify regions under selection.

```
> plot(FST[,1],type="l",xlab="SNP",
+    ylab="Fst",col="gray")
```

This looks rather messy (the grey lines in Fig. 4.8). There is a lot of noise in the estimates. Since there is high LD we could try to smooth the F_{ST} values to try to identify clearer patterns across the chromosome. The *runmed* function can be used for this; it *dampens* the individual fluctuations by computing median values across a moving window.

```
> smoothed=runmed(FST[,1],k=101,endrule="constant")
> lines(smoothed,type="l",col="red",lwd=6)
```

The k parameter sets the window size for the running median (it has to be an *odd* number). Small windows will have little *shrinkage* while larger windows will have stronger *damping* effects—on the extreme sides, if $k = 1$ the values will be the same as the original data; with $k = 7,763$ (number of SNP) the results will almost be a straight line through the median of the whole data. Try changing the values of k to get a feeling for this. Here $k = 101$ was used, which is a reasonable value for this data based on the distance between markers and the level of LD in these populations; there is no exact way to calculate the ideal number for k: some knowledge about LD and map distances can help but it is also worthwhile to try different values. Signals at the beginning and end of the chromosome are not reliable due to the smoothing method, the argument *endrule* was set to *constant* to avoid signals on the tails. The second line of code adds the results of *runmed* to the plot (Fig. 4.8).

Now, in Fig. 4.8 with the smoothed values plotted we can see a much clearer signal that is higher than the others (around SNP 3,500). Highly divergent loci or genomic regions between populations have high F_{ST} values and are potentially associated with either natural or artificial selection—the higher the value the stronger is the evidence for selection (differentiation). To this point we have relied only on visual identification of SOS regions but these should be tested more

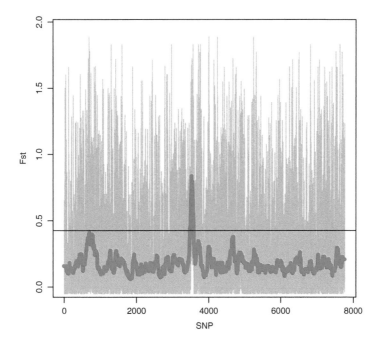

Fig. 4.8 Genome-wide plot of FST and smoothed FST (*thick line*). Notice how the smoothed FST highlights a region under potential selection around SNP 3,500—it is less clear just from the individual FST values due to the strong fluctuations from SNP to SNP. The *horizontal line* shows the threshold of significance

formally. A simple albeit suboptimal approach is to calculate the average and standard deviation of the F_{ST} values for each population, and identify the regions that have values greater than the average plus three standard deviations (around 1 % top values). Better but computationally more demanding is to use a resampling approach and obtain *p*-values from the empirical distributions. Let's do the easy one here:

```
> FSTm=mean(smoothed)
> FSTsd=sd(smoothed)
> sig=FSTm+3*FSTsd
> abline(h=sig)
> print(sig)

[1] 0.4249352
```

We used the smoothed values to calculate the significance threshold (0.425) but could also have used the raw F_{ST} values (note that there is too much noise and the dataset is rather small). The function *abline* was used to place the significance threshold on the plot and complete Fig. 4.8. There is only one region that shows a strong SOS—which is good, this is the region where I did some data *tweaking* of the genotypes for illustration purposes. If this was real data, the next step would be to identify the significant region in terms of chromosomal position and search the databases for biological inferences (see Chap. 6 for these next steps).

4.6.2 Other Population Estimates

SNP data is very well suited for population genetics studies: diversity, population structure, selection, phylogenetics… We have already seen some of the common population metrics but in this section we will focus on using available R packages to calculate parameters—the objective is to briefly overview some useful functions for common parameters. To start, we will calculate F_{ST} for the same data again using the *pegas* package.

```
> library(pegas)
> M=t(sos)
> M[which(M==0)]="0/0"
> M[which(M==1)]="1/0"
> M[which(M==2)]="1/1"
> M=as.data.frame(M)
> M=as.loci(M)
> M

Allelic data frame: 75 individuals
                     7763 loci

> FST2=Fst(M,pop=rownames(M))
> head(FST2)

            Fit         Fst           Fis
snp1  0.6192893  0.4744078   0.27565392
snp2  0.6684625  0.6831399  -0.04632153
snp3  0.2963940  0.3275432  -0.04632153
snp4  0.2009834  0.1139111   0.09826590
snp5  0.3479549  0.2960877   0.07368421
snp6  0.1281274  0.1881476  -0.07392996

> cor(FST2[,2],rowMeans(FST))

[1] 0.9952647
```

First we had to do some data format conversion—*pegas* uses a *loci* structure. It can import data in many formats, unfortunately not the one we had here. All we did was replace the genotypes from 0, 1 and 2 to *0/0*, *1/0* and *1/1*. Note that it is a matrix but it was treated as a vector for the replacement—this is quite handy: R can treat a matrix as a vector for operations (much faster than using, e.g., *apply*). Two other points: for *pegas* the data has to be in *sample by SNP* format (hence the transpose), also only accepts *data.frames*. The function *as.loci* converts the data into a format that can be used with *pegas's* functions. The function *Fst* calculates F_{ST} as well as F_{IT} and F_{IS}; note that these are average values for the three populations. The correlation with the F_{ST} values we calculated before is very high (0.995). The package will also calculate other parameters such as Hardy–Weinberg equilibrium (*hw.test(M)*), R^2 (*R2.test(M)*), linkage disequilibrium (*LD2(M)*), Tajima's D (*tajima.test(M)*), and others. Another useful package is *hierfstat*. One of the functions (*basic.stats*) will calculate the most common parameters all at once. These include allele frequencies in each population, observed heterozygosities and gene diversities, F_{ST}, F_{IS}, and others. Again requires some data shuffling to fit the package requirements.

```
> library(hierfstat)
> M2=data.frame(pop=colnames(sos),t(sos))
> basStats=basic.stats(M2)
> names(basStats)

[1] "n.ind.samp" "pop.freq" "Ho" "Hs"
[5] "Fis" "perloc" "overall"

> # observed heterozygosity
> head(basStats$Ho)

      angus brahman hanwoo
snp1   0.04    0.40   0.84
snp2   1.00    0.00   0.68
snp3   1.00    1.00   0.84
snp4   0.56    0.16   0.68
snp5   0.96    0.96   0.68
snp6   0.28    0.12   0.68
```

There are many other packages (some listed at the end of the chapter) and there is some redundancy across packages. The downside is that each one expects data in a certain way, but there are also many useful functions between them; in combination they cover most of the basic metrics used in population genetics. Check the documentation of the packages for more information.

4.6.3 Genetic Distances

To close this section, a brief look at genetic distances between populations. Genetic distances help us understand the evolutionary relationships between individuals, populations, species; how they share a particular lineage and descend from a common ancestor. Readers interested in phylogenetics will find most of what they need in the *ape* package; the excellent book by Paradis [84] is also a must for the field. Here we will focus on a few simple examples using SNP data to estimate distances in populations of a same species. This is a little different from phylogenetic analyses that use data from different species. With multiple species there is usually a multiple sequence alignment (*MSA*) step that has to be performed prior to the phylogenetic analysis itself. A sequence alignment (with DNA, RNA, or even protein data) tries to identify regions of similarity between sequences and then aligns them in such a way as to reflect evolutionary relationships between the sequences. This is based on a *model of molecular evolution* and an algorithm is used to find the optimal alignment (or almost optimal—these methods are usually heuristic due to computational constraints). A common program for multiple sequence alignment is *ClustalW*; a somewhat *incidental* example of MSA is shown in Chap. 7. Once the sequences have been aligned, different measures of distances can be calculated and these can then be used to build relationship trees (phylogenetic trees—a branching diagram that shows the inferred *evolutionary* relationships between the organisms sampled). A brief example with mitochondrial sequence data is shown in Chap. 7. With samples from the same species, data can be aligned against a reference assembly or, in the case of SNP chips, the data is already matched up with the SNP from the array; so there is no need for a MSA. Two very useful packages to handle sequence data are *Biostrings* and *SeqinR*.

Here we will continue working with the genotypes from the three cattle breeds (the *sosData.rds* file). First, the *StAMPP* package will be used to estimate pairwise F_{ST} between the three breeds. Up to this point we have only looked at F_{ST} per SNP, but we can also estimate average F_{ST} across all SNP and how different each population is from the others.

```
> read in the data again, if needed
> sos=readRDS("chapter4/sosData.rds")

> library(StAMPP)

> # some format fixing up
> M=t(sos)
> M[M==0]="AA"
> M[M==1]="AB"
> M[M==2]="BB"
> M=cbind(sample=1:nrow(M), pop=rownames(M),ploidy=2,
+    format="BiA",M)
> M=as.data.frame(M)

> M=stamppConvert(M,type="r")

> M[1:6,1:6]

    sample     pop pop.num ploidy format snp1
1        1  hanwoo       1      1    BiA  0.5
2        2  hanwoo       1      1    BiA  0.5
3        3  hanwoo       1      1    BiA  0.0
4        4  hanwoo       1      1    BiA  1.0
```

| 5 | 5 hanwoo | 1 | 1 | BiA | 0.0 |
| 6 | 6 hanwoo | 1 | 1 | BiA | 0.5 |

Once again, we had to organize the data to the format requirements of the package. The function requires the data to be a *data.frame* and the first column should have sample identifiers (here we just used numeric *invented* ids), the second column should contain population IDs, the third column is numeric with the *ploidy* level (i.e., diploid = 2), the fourth column is a code for the format of the genotype data (*BiA*—biallelic AB format), and finally the SNP. The *stamppConvert* function converts the *AA*, *AB*, and *BB* to *0.0*, *0.5*, and *1.0* and adds an extra *pop.num* column (numeric levels of the populations).

The *stamppFst* function calculates pairwise F_{ST} values and also confidence intervals and p-values between the populations. The parameters in the function are *nboots*—number of bootstraps across loci for the confidence intervals and p-values; *percent*—the percentile for the confidence interval and *nclusters*—is the number of cores to use for the calculations. Note that this is quite slow with large datasets.

```
> # global pairwise FST
> FST=stamppFst(M,nboots=200,percent=95,nclusters=8)
> FST$Fsts
```

```
$Fsts
            hanwoo     angus brahman
hanwoo          NA        NA      NA
angus    0.3747484        NA      NA
brahman  0.4505593  0.530454      NA
```

The number of breeds is sufficiently small to visually identify that the brahman are more removed from angus and hanwoo; the largest difference is between angus and brahman while the smallest is between angus and hanwoo. The results are sensible since the hanwoo and angus are *Bos taurus* animals and the brahman are *Bos indicus*. The first two are closely related whilst the common ancestor between *Taurine* and *Indicine* cattle is much more ancestral. The F_{ST} values are much higher than would be expected in real data—this is largely due to the high *LD* between markers in each population (recall this is only a small part of a chromosome and there was some artificial tweaking of data). Linkage disequilibrium is the non-random association of alleles at two or more loci that have been co-inherited by descent from the same ancestral chromosome. We can estimate *LD* in this data with the *snpStats* package. This time it is simple to convert our data into a *snpMatrix* object for the package and the function *ld* can be used to calculate LD between pairs of SNP.

```
> library(snpStats)

> # convert to SnpMatrix object
> # plus one is because SnpMatrix
```

```
> # codes genotypes as 1,2,3
> # and 0 is for missing

> M=new("SnpMatrix",(t(sos)+1))

> # calculate LD in hanwoo
> ldHan=ld(M[1:25,],depth=1000,stats="D.prime")

> # LD across all individuals
> ldAll=ld(M,depth=1000,stats="D.prime")

# plot LD
> cols=colorRampPalette(c("yellow", "red"))(10)
> image(ldHan[1:1000,1:1000], lwd=0, cuts=9,
+   col.regions=cols, colorkey=TRUE)
> windows()
> image(ldAll[1:1000,1:1000], lwd=0, cuts=9,
+   col.regions=cols, colorkey=TRUE)
```

The arguments to *ld* are *depth* (how many pairwise combinations to run, here up to 1,000 SNP apart from each other) and *stats* (the measure of LD, either the common *D.prime* or *R.squared* or one of another four—details in the documentation of the package). Another handy argument is *symmetric* (*TRUE* or *FALSE*, if true returns a symmetric matrix, which is useful for other functions). Results were then plotted using *image* and a gradient of colors between yellow and red—similar to a *heatmap*. Figure 4.9 shows the pairwise LD for the first 1,000 SNP in hanwoo and the LD for all individuals is shown in Fig. 4.10. Note how very high levels of LD extend throughout the whole region in hanwoo but, as expected, are much lower with distinct populations. A note on the *snpStats* package, it was designed to handle quite large datasets and will also compute many population parameters (e.g., F_{ST} with the *Fst* function—much faster than the version implemented in *pegas*).

Now we are ready to build a tree of genetic distances for the individuals in the three populations. A simple method is *allele sharing*—it is reasonably fast and usually works well with populations of the same species. For every SNP of every individual, the absolute differences in genotypes against all other individuals are taken (genotypes coded as 0, 1, 2), and then the mean of the differences across all SNP is calculated for each individual. In other words, the allele sharing for two individuals is the average of the absolute difference between all their SNP. Let's first calculate for the first two individuals in our population:

```
> M=readRDS("chapter4/sosData.rds")
> mean(abs(M[,1]-M[,2]))

[1] 0.436558
```

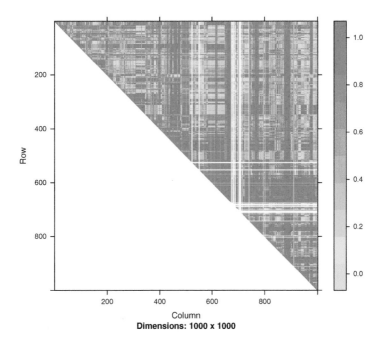

Fig. 4.9 Extent of linkage disequilibrium in hanwoo calculated as D'. There is strong LD along large chromosomal segments. *White areas* are SNP with no variation and for which D' could not be calculated. The figure shows all pairwise LD for up to 1,000 SNP and the first 1,000 SNP are shown on the plot

Now for the whole population:

```
> # allele sharing
> allshare=matrix(NA,ncol(M),ncol(M))
> for(i in 1:ncol(M))
+ {
+   hold = abs(M - M[,i])
+   allshare[,i] = colMeans(hold,na.rm=T)
+ }

> # names of breeds
> colnames(allshare)=colnames(M)
> rownames(allshare)=colnames(M)

> library(ape)
> tree=as.phylo(hclust(as.dist(allshare),
+   method="ward.D2"))

> # just for plotting purposes
> # to assign colours to samples in a population
> pop=as.factor(colnames(M))
```

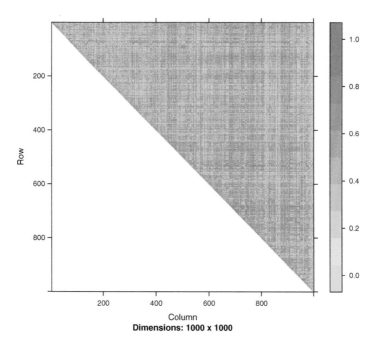

Fig. 4.10 Extent of linkage disequilibrium in three unrelated populations calculated as D'. There is weak LD across chromosomal segments. The figure shows all pairwise LD for up to 1,000 SNP and the first 1,000 SNP are shown on the plot

```
> cols=c("black","blue","red")

> plot(tree,type="cladogram",edge.color="gray",
+   direction="r",  cex=0.7,adj=0,
+   tip.color=cols[as.numeric(pop)])
> legend("topleft",levels(pop),fil=cols,cex=0.8)
```

The *ape* library has the *as.phylo* function which is convenient to convert a distance matrix into a phylogenetic tree for use with the plotting functions of *ape*. The tree is shown in Fig. 4.11, results are consistent with our expectations and all three breeds cluster separately from each other (with the individuals in each breed also fully within their own groups). The function *as.dist* was used to convert the allele sharing matrix into a *distance* object; this does not change the values in the matrix, just the representation. Be careful not to use the *dist* function—that would return a distance matrix computed using one of the distance measures implemented in the function (we would end up with a distance of distances). The default distance measure is *euclidean* but there are another five options (*maximum, manhattan, canberra, binary,* and *minkowski*) which can be changed with the *method* argument. As an exercise try some of the distance measures straight on the matrix of genotypes and then build a tree. The function *hclust* performs hierarchical clustering of a

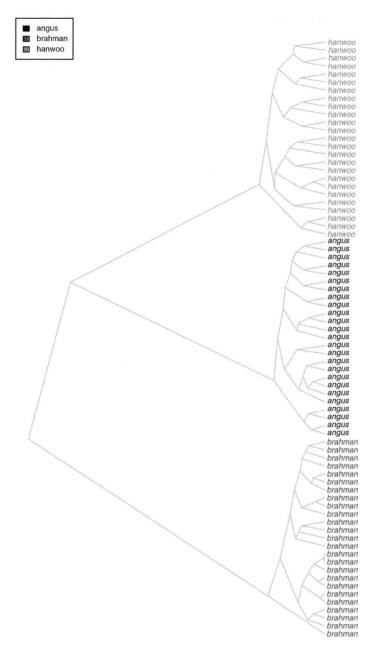

Fig. 4.11 Genetic distance tree for all individuals in three cattle populations built with *allele sharing*

distance matrix (needs a *dist* object). The function implements various clustering algorithms (same as with *dist*, use *method* argument to change); the most common ones are *complete* which is the default and finds more similar clusters, the *average* method is the same as UPGMA and Ward's method which tends to shrink the variance within groups and inflate between groups (two implementations: *ward.D* and *ward.D2*). The distance and clustering methods have a considerable effect on the final tree, it is worthwhile exploring different options to become familiar with the properties of each method. The plotting functions for a *phylo* object are comprehensive and well detailed in [84], but the most important one is the type of tree to build (argument *type*), any one of *phylogram*, *cladogram* (used in the example), *fan*, *unrooted*, or *radial*.

4.6.3.1 *GRM* and Genetic Diversity

The genomic relationship matrix (*GRM*) is also useful for estimating genetic distances and evaluating the genetic variability between populations. The *GRM* can be built as before with

```
> M=M-1 # recode matrix as -1, 0, 1
> p=rowMeans(M,na.rm=T)/2
> P=2*(p-0.5)
> Z=M-P
> ZtZ = t(Z) %*% Z
> d=2*sum(p*(1-p))
> G=ZtZ/d

> rownames(G)=pop
> colnames(G)=pop
```

And it can be treated as a matrix of genetic distances between all individuals. The *GRM* can be used in the same way as the allele sharing matrix to build a phylogenetic tree (simply replace *allshare* with *G* in the code above); or alternatively a *heatmap* is quite informative.

```
> heatmap(G,symm=T,col=gray.colors(16,start=0,end=1),
+   RowSideColors=cols[as.numeric(pop)],
+   ColSideColors=cols[as.numeric(pop)])

> legend("topleft",levels(pop),
+   fil=cols[1:length(levels(pop))])
```

The heatmap for the three populations is shown in Fig. 4.12. The arguments *RowSideColors* and *ColSideColors* were used to color code the sides of the *heatmap* based on the population of origin of each individual. The results are the same as obtained with the allele sharing method used above; all individuals cluster within

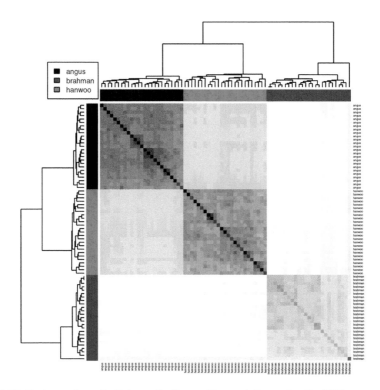

Fig. 4.12 Heatmap of genetic distances for three cattle populations using the *GRM*

their own breed groups; the angus and hanwoo are more related with the brahman being the most dissimilar. The block structures of the *heatmap* make it easy to identify relationship groups—notice how even within a breed there are individuals more and less related to each other (the smaller blocks along the diagonal). Here the color gradient used had 16 shades of gray (*gray.colors(16,start=0,end=1)*), use less colors for stronger (and coarser) differentiation or more colors to pick up subtle details (change the number of colors to see how it affects the block structures).

Lastly, singular value decomposition (*SVD*) of the *GRM* matrix can be used to evaluate diversity in the populations. SVD is the main algorithm behind principal component analysis (*PCA*). In general terms PCA is a method that defines a new set of variables that capture the variation in the data and *project* it into new uncorrelated variables. Whilst the original data can have correlations within it (which makes some points superfluous), the principal components are uncorrelated (orthogonal). There can be as many components as data points and components are defined in such a way as to ensure that each successive component explains less variation than the previous one (i.e., *PC1* should be the most explanatory, then *PC2*...). A note on PCA is that it is sensitive to the original values—these should be mean centered and preferably normalized. Components can be used for prediction purposes, dimensionality reduction, account for fixed effects in the data (e.g., population structure in a GWAS); identification of factors in multivariate analysis and, in our case, it can be used to identify the sources of genetic variation in the populations. Chapter 4 from [116] has a very good description of PCA and the underlying numerics.

```
> G=G-mean(G) # mean centering

> SVD=svd(G)

> plot(SVD$v[,1],SVD$v[,2],cex.main=0.9,
+   main="Principal components",pch=16,
+   xlab="PC1",ylab="PC2",col=cols[pop])
> legend("bottomright",levels(pop),
+   fil=cols[1:length(levels(pop))],cex=0.8)
```

It is quite easy to do an SVD in R using the *svd* function, just remember to scale the variables appropriately (other useful functions are *irlba* for very large datasets, *eigen* and *princomp*). The *svd* function decomposes a matrix (here the *GRM*) and returns a list with the three components: *u* (left singular vectors), *d* (singular values), and *v* (right singular vectors), such that $X = UDV'$. In PCA terms UD are the scores and V the loadings. The variance explained by each of the components can be calculated with:

```
> # variances of components
> variances=SVD$d^2/(ncol(G)-1)
> variances=round(100*variances/sum(variances),3)
> variances[1:2]
```

```
[1]  72.571  22.082
```

Figure 4.13 shows the plot of the first two principal components from the SVD. Most of the variation in this dataset is already captured by the first two components. *PC1* explains 72.57 % of the variance and *PC2* an additional 22.08 %. This again fits in with previous results, with the brahman being the most distant population (*PC1* splits the data into taurine and indicine cattle) and almost 73 % of the variation is due to the difference between these two subspecies. *PC2* then splits the European and Asian cattle into two groups (another 23 % of the variation). The difference between individuals within a population is small at around 5 %. Notice that three angus and one brahman (particularly the latter) do not cluster as tightly as the other individuals; this could be due to, e.g., crossing with other breeds (could also simply be individuals less related to the others or even a genotyping problem).

This concludes our brief exploration of population genetics. We overviewed the main metrics used to understand population structure, variability, and relationships. There is a wide range of R packages and functions that make it relatively simple to perform these analyses, possibly the main difficulty is getting data into the different formats expected by these packages. These metrics are not only relevant to understand populations but are also useful for GWAS or genomic prediction. Measures of LD can inform sample sizes needed for identification of QTL or

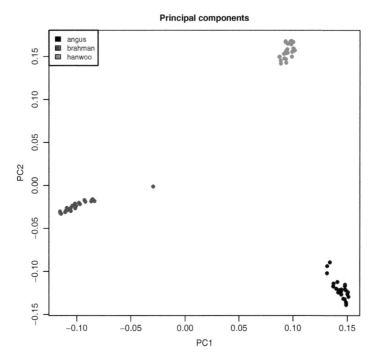

Fig. 4.13 Plot of singular value decomposition of the *GRM* showing the first two PCs

the power to detect effects; population stratification (e.g., breeds) should be used to account for different genetic backgrounds; signatures of selection can be used as another approach to identify genetic association with a trait; and even for quality control purposes: a plot of principal components is an easy way to identify mislabelled samples (e.g., wrong breed assignment) or outliers (such as one of the brahman in Fig. 4.13). As a rule, when working on association or prediction studies, some of these metrics should be part of the workflow—they will provide a better understanding of the data and help drive experimental questions and analytical methods.

In relation to the three cattle breeds, we now know that there is strong linkage in each population which suggests a small effective population size (at least in this data); we found a signature of selection in hanwoo (unfortunately a *fabricated* one); we used various measures of diversity to characterize the populations (F_{ST}, *allele sharing*, *GRM*, *SVD*) and found that the largest differentiation is between taurine (angus and hanwoo) and indicine (brahman) cattle, which fits what is known about breed formation and domestication—these two groups belong to independent domestication events and have around 4 million years of isolation between them. Note however that brahman have a taurine background (on average around 8 %) and are not pure indicine (if we had pure *Bos indicus* cattle we would see that the brahman come closer to the taurine in, e.g., a PCA plot). Over 75 % of the variation is due to the difference between these two subspecies (*Bos primigenius taurus* and *Bos primigenius indicus*), and it would be even more with pure indicine cattle. Another 22 % of the variation is due to an *East versus West* gradient of dispersal. Taurine cattle are assumed to have been domesticated in the Near East (Fertile Crescent) around 10,000 years ago and then dispersed into Europe (angus in this example) and Asia (hanwoo). There is only another 5 % of variation within breeds which matches the high LD results. In a real diversity study it would be necessary to take into account the gene flow between these populations, livestock in general has not evolved independently, there has always been some level of migration due to human movements and trade; there is also over representation of some genetic lines in the various breeds that further complicate matters. For example the brahman outlier in Fig. 4.13 could be a cross between brahman and taurine cattle or a composite animal (e.g., droughtmaster—cross of brahman and European cattle).

4.7 Parentage Testing

SNP can be used for parentage verification (identification of parent–offspring relations). In humans, accurate determination of paternity has a very obvious societal value; but it is also relevant for livestock production; production performances need to be linked back to the correct individuals and families so that estimates of breeding values are correct and inbreeding levels are monitored. Pedigree information is usually used for these purposes, however problems can occur due to missing data, human error or even willful forgery. In any of these cases, a DNA-based parentage test can clarify the ancestry and include (or exclude) a *putative* parent of an offspring. In livestock, SNP are rapidly replacing microsatellites as the marker of choice for parentage testing due to their ease of automation, lower genotyping cost per marker and standardization between different laboratories. From an information content perspective SNP are only bi-allelic and more of them are needed to obtain the same level of information contained in the highly polymorphic microsatellites. As a general approximation, between 40 and 100 SNP are equivalent to between 14 and 20 microsatellites.

Acceptance or rejection of a parentage test can be based on a probabilistic model conditioned on allelic frequencies of the population, or simply by testing for Mendelian inconsistencies between a putative parent–offspring pair. A Mendelian inconsistency occurs when an individual shows an allele that could not have been received from its parent. For example the parent has a genotype *AA* for a SNP at a given locus and the offspring has a *BB*; this should not occur as the offspring would have to have inherited one of the *A* alleles from its parent (the genotype would have to be either *AA* or *AB*, depending on what allele it received from the other parent). Such a Mendelian inconsistency can only occur if there was a mutation at that locus (extremely rare event), a genotyping error (more common) or if they are not a parent–offspring pair. So, true parent–offspring relations should not have any Mendelian inconsistencies but these do occur due to genotyping errors and, in practice, a 1% mismatch can be adopted as an acceptable error rate.

In this section we will look at how to perform a parentage test based on Mendelian inconsistencies, what genomic relationships look like, and the effect of the number of SNP on the accuracy of parentage testing. This will be based on *opposing homozygotes* which occur if one individual is homozygous for one allele and the other individual is homozygous for the other allele (as in the example above). All we have to do is count the number of opposing homozygotes (*OP*) between every pair of individuals and, those that have a count of zero, are parent–offspring pairs—except for the genotyping errors which bring the OP counts up (mutation is negligible). We will use a *highly adapted* subset of SNP with 20,000 SNP and 312 individuals in family structures (the *parData.rds* file). There are 12 sires (fathers) and 108 dams (mothers), plus 96 full-sibs (8 per sire/dam) and another 96 half-sibs (also 8 per sire). In the full-sibs the same dams were used in each family (12) and in the half-sibs all dams were different (96). Sires and dams are unrelated. We also have an accurate pedigree (*pedigree.txt*) file to work with. Start by reading in the data with

```
> M=readRDS("chapter4/parData.rds")
> dim(M)

[1] 20000    312

> ped=read.table("chapter4/pedigree.txt",
+      header=T,sep="\t")
> head(ped)

  id sire dam group
1  1    0   0     0
2  2    0   0     0
3  3    0   0     0
4  4    0   0     0
5  5    0   0     0
6  6    0   0     0
```

Genotypes are already in the same order as the pedigree. In the pedigree (*ped*) the sires and dams have *0* for their sires and dams (common notation for unknown or base population in a pedigree). The last column group splits the individuals into *0 = sire, 1 = dam, 2 = full-sibs* and *3 = half-sibs*.

To build the matrix of opposing homozygotes (counts of OP for every pair of individuals) we can use R's matrix capabilities.

```
> # OH matrix
> down=up=matrix(0,nrow(M),ncol(M))
> up[M==2]=1
> down[M==0]=1
> op=t(up)%*%down
```

```
> op=t(op)+op
> rm(up,down)
```

In the code above, two empty square matrices (*number of individuals by number of individuals*) were created—matrices *up* and *down*—note that both were created with the same command line. Then, in the positions where the genotypes are homozygous 2, we put a *1* in the *up* matrix and likewise, when genotypes are *0* we put a *1* in the *down* matrix. These are of course just incidence matrices for the occurrence of the two homozygous genotypes. Then it is just a matter of multiplying the matrices to get the partial counts (upper diagonal incidence of 2 and 0; lower diagonal incidence of 0 and 2); and adding the top and bottom part of the matrix to get final (and symmetric) counts.

```
> op[1:5,1:5]

      [,1]  [,2]  [,3]  [,4]  [,5]
[1,]     0  1267  1341  1327  1353
[2,]  1267     0  1301  1347  1326
[3,]  1341  1301     0  1293  1333
[4,]  1327  1347  1293     0  1284
[5,]  1353  1326  1333  1284     0
```

The same values, for the first pair, can be obtained with

```
> top=length(which(M[,1]==0 & M[,2]==2))
> top

[1] 662

> bot=length(which(M[,1]==2 & M[,2]==0))
> bot

[1] 605

> sum(top+bot)

[1] 1267
```

Let's take a first look at the data with a *heatmap* of the OP matrix, the same way as before with the genomic relationship matrix for the three cattle populations. The first few lines of code are simply to build a vector for color coding the sides of the *heatmap*. The function *getcol* in the package *made4* is useful to get a range of colors that are clearly distinguishable from one another (up to a maximum of 21).

```
> # build levels for plotting
> groups=ped$sire
> groups[1:12]="sire"
> groups[13:120]="dam"
```

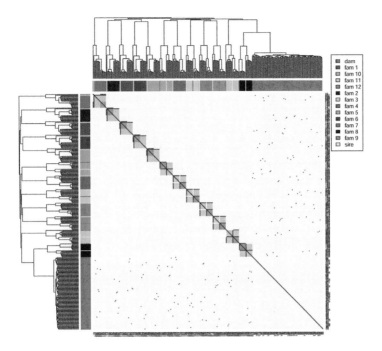

Fig. 4.14 Heatmap of family relationships based on matrix of opposing homozygotes

```
> groups[121:312]=paste("fam",groups[121:312])
> groups=as.factor(groups)

> # get colours for plot
> library(made4)
> cols=getcol(length(levels(groups)))

> heatmap(op,symm=T,col=gray.colors(32,start=0,end=1),
+   RowSideColors=cols[as.numeric(groups)],
+   ColSideColors=cols[as.numeric(groups)])
> legend("topleft",levels(groups),
+   fil=cols[1:length(levels(groups))],cex=1.3)
```

Figure 4.14 shows the *heatmap* for the matrix of opposing homozygotes. The figure is not intuitive at first glance but if it is broken down into its various components and related back to the pedigree structure it becomes clearer. The diagonal is the similarity of the individual with itself (dark color), across the diagonal there are 12 blocks—these are the 12 sires and their offspring (16 for each sire, 8 full-sibs and 8 half-sibs). A quarter of each block is darker, these are the full-sibs and the relationship is stronger because they are more related to each other (0.5) than half-sibs (0.25). The full and half-sibs have the same relationship to each other (sire in common). Still within the blocks, each one has either a strong dark border along the sides of the block (e.g., 1–6 from the top) or cutting across it (e.g., block 6 from the top)—these are the sires of each family. Apart from genotyping errors there should be no OP between sire and offspring, they should show the same *relatedness* as an individual to itself (remember this is the number of opposing homozygotes, not *genetic* relationship). The *heatmap* assigns the order only based on clustering; the sires can be either on the borders or inside the family block. There is still another structure inside the blocks which is similar to the sires but now half white, half black. These are the dams of the full-sibs, they have no OP with their eight offspring (note how the dark part is always around the full-sib blocks) but they are unrelated to the half-sibs (hence the white). Then there is a large region with no relations except for a few dark spots: these are the dams of the half-sibs, note that there are eight per family. The *heatmap* clearly identifies all features in the data, alternatively a *GRM* could also be used and results should be reasonably similar (although more error prone). In practice this can be used to check and fix pedigree errors—offspring with the wrong color in a family group can be easily identified (as an exercise, assign one of the offspring to another sire and redo the heatmap).

Returning to the parentage testing problem, a plot of the sorted counts of opposing homozygotes (Fig. 4.15) makes it easy to identify the different relationships. Visually it can be seen that true parent–offspring relationships have less than 200 OP, full-sib groups around 400 OP, half-sibs 700 OP, and unrelated pairs upwards of 1,200 OP. Note how there always is some level of genotyping error (none of the parent–offspring relations have 0 OP). Now, this is a good dataset, and all levels of relatedness are well defined. In practice there can be uncertainty about some groupings (there is no gap between one and another—the plot is uninterrupted) but not in the case of parent–offspring, these are always well defined provided the SNP panel is reasonably large (further discussed later on).

```
> plot(sort(op[upper.tri(op)]),main="sorted op counts",
+   ylab="number of op",xlab="",pch=20)
> abline(h=200)
```

The next step is to identify parent–offspring pairs. Figure 4.15 already showed that any pair with less than 200 OP will be a correct parent–offspring pair. We know the first 12 animals are sires, their offspring can be identified with

```
> # build sire-offspring pairs
> relations=NULL
> for (i in 1:12)
+ {
+   off=which(op[i,]<200)
+   off=off[- which(off==i)]
+   relations=rbind(relations,
+      data.frame(sire=rep(ped$id[i],length(off)),
+      offspring=ped$id[off]))
+ }
> head(relations)
```

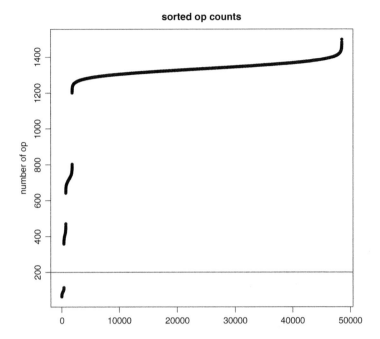

Fig. 4.15 Sorted counts of opposing homozygotes for all pairwise comparisons

```
   sire offspring
1    1       121
2    1       122
3    1       123
4    1       124
5    1       125
6    1       126
```

For each sire the entries with less than 200 OP were found (had to remove the entry of the individual with itself, hence the second line in the loop) and then matched to the *ids* in the pedigree (note that pedigree information was never used, we *assembled it de novo*). This was simple since we knew the sires, if not, all pairs would have to be tested—not complicated, just more computationally intensive.

As a last exercise, we will have a look at how many SNP are needed for parentage testing. For this, a smaller dataset will be used—only the sires and their half-sibs.

```
> M=M[,which(ped$group==0 | ped$group==3)]
> ped=ped[which(ped$group==0 | ped$group==3),]
```

The strategy is to select random SNP in increasing numbers and measure how many are correctly assigned and how many are incorrectly assigned with different panel sizes. The threshold for correct/incorrect will be 1 % (i.e., the parentage assignment is accepted up to a maximum of one percent OP mismatches). The code is mostly about pulling together what has been done so far (note that it is quite specific to this particular dataset).

```
> # random SNP in intervals of 100
> SNPinterval=seq(100,5000,100)
> # threshold
> tol = 0.01
> # number of sires
> numSires=12
> # offspring per sire
> numOff=8

> # store results for
> # different number of SNP
> results=NULL

> for (i in 1:length(SNPinterval))
+ {
+    # number of SNP to sample
+    numSNP=SNPinterval[i]
+    # sample random SNP
+    rM=M[sample(nrow(M),numSNP),]

+    # OH matrix of subset
+    down=up=matrix(0,nrow(rM),ncol(rM))
+    up[rM==2]=1
+    down[rM==0]=1
+    op=t(up)%*%down
+    op=t(op)+op
+    rm(up,down)

+    # hold results for each sire
+    tot=numeric(numSires)
+    correct=numeric(numSires)
+    incorrect=numeric(numSires)
+    mis=numeric(numSires)

+    for (j in 1:numSires) # loop over sires
+    {
+       # accept 1% mismatch
+       off=which(op[j,] <= (numSNP*tol))
+       off=off[- which(off==j)]
+
+       # total accepted
+       tot[j]=length(off)
```

```
+        # number correct
+        correct[j]=length(intersect
+        (ped$id[off],ped$id[which(ped$sire==j)]))
+        # number incorrect
+        incorrect[j]=tot[j]-correct[j]
+        # number missing
+        mis[j]=numOff-correct[j]
+      }


+      results=rbind(results,data.frame(
+        total=mean(tot),
+        correct=mean(correct),
+        incorrect=mean(incorrect),
+        missing=mean(mis)
+        ))
+ }
```

The structure of the code is to randomly sample SNP panels of different sizes to use for parentage testing, build an OP matrix with each panel, infer offspring for each sire, and compare results with the real pedigree to evaluate efficacy of the panels. Panel sizes range between 100 and 5,000 in 100 SNP increments (these are held in the *SNPinterval* variable created with the *seq* function). The *sample* function is used to select SNP for the random panel (a convenient function to sample unique *indexes* that can be used to randomly subset data). The matrix of OP is then built for the subset of genotypes (*rM* variable) as before. Then the code loops over each of the sires and assigns offspring to them; this is also the same as we did before but now instead of the arbitrary cutoff of 200 OP that was visually chosen, now the threshold is based on a 1 % tolerance for genotyping errors, i.e., for 200 SNP up to 2 OP are accepted as a true parent–offspring relation (this is defined in the variable *tol*). The rest are just metrics to evaluate the panels: *tot*—total number of offspring accepted (not necessarily all correct); *correct*—number of correct offspring identified (note the use of the *intersect* function to match inferred relations with true relations from the pedigree); *incorrect*—number of offspring incorrectly identified (incorrectly assigned to a sire); and *mis*—missing, how many offspring were not assigned to a sire. For each panel these values are averaged out across all sires and placed in the *results* variable (stores average for all sires for each panel size).

To make it easier to visualize, *results* can be scaled into percentages (note that results will vary between runs)

```
> results=results/numOff*100
> head(results)
```

	total	correct	incorrect	missing
1	96.87500	87.50000	9.375	12.500000
2	94.79167	94.79167	0.000	5.208333
3	98.95833	98.95833	0.000	1.041667

```
4   95.83333   95.83333      0.000   4.166667
5   93.75000   93.75000      0.000   6.250000
6  100.00000  100.00000      0.000   0.000000
```

If the columns *total* and *correct* have the same value, there will be no incorrect assignments; *total* can be more than 100%, meaning that more offspring were assigned than there really are. The number of incorrect assignments is already zero from 200 SNP onwards (in this particular run, but it is quite consistent), this indicates that the threshold is quite stringent (note that *missing* starts higher and gradually drops down). If the value in *tol* is increased to 0.02 or 0.03 (2–3%) there will be less *missing* but more *incorrect*—a percentage based cutoff is a compromise between (true/false) positives and negatives (try different values and compare results). Figure 4.16 shows how the proportion of correct parentage assignment increases with the number of SNP in the panel. We ran the example only up to 5,000 SNP and not all 20,000; the figure makes it clear that with more than 1,000 SNP almost any panel is 100% accurate (and after 1,500 all are). Of course, since we are working with random SNP we should repeat the sampling to get a better handle on the variation due to sampling—but for illustration purposes this suffices; and in reality this does not change much in terms of the trend—larger gains up to 1,000 SNP, from there on it makes hardly any difference (and most SNP sets will be perfect for parentage assignment).

```
> plot(SNPinterval,results[,2],pch=20,
+     xlab="number of SNP",
+     ylab="percentage",
+     main="percentage of correct parentage assignment")
> abline(v=1000)
```

A final note on parentage testing. *True* parentage panels use between 100 and 200 SNP, sometimes up to 400, they are carefully selected (based on allele frequencies, chromosomal distribution, high genotyping success. . .) and work better than random SNP; it is possible to obtain good results with a smaller panel than what we observed here. However, the best way to increase accuracy of a parentage panel is to simply add more SNP (at a higher genotyping cost, of course).

4.8 Useful R Books and Packages

- Ridge regression BLUP and *gBLUP:biRR, BLR, rrBLUP* and *pedigree*. The book *Genome-Wide Association Studies and Genomic Prediction* [46] covers most methods for genomic prediction and many examples are in R.
- Estimation of population genetic parameters: *genetics, adegenet, pegas, genAbel, hierfstat, PopGenome* (suitable for sequence data) and *snpStats*.
- Phylogenetics: *ape* is the broadest package for phylogenetic inferences, the book *Analysis of Phylogenetics and Evolution with R* [84] is a must read for

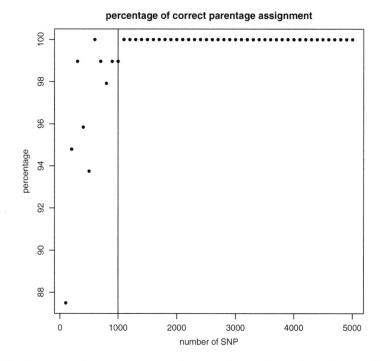

Fig. 4.16 Percentage of correct parentage assignment using varying numbers of random SNP

using R in this field. There are many other packages listed in the CRAN task view: *Phylogenetics, Especially Comparative Methods*. Other useful packages are *DECIPHER* for alignments; *Biostrings* and *SeqinR* for handling sequence data.

- Parentage: the *hsphase* package can calculate the OP matrix and has some useful plotting functions, the *pedigree* package has useful functions to build and manipulate pedigrees (it also calculates inbreeding from a pedigree), other useful packages are *kinship2, pedigreemm* and *GeneticsPed*.

Chapter 5
Gene Expression Analysis

In this chapter we will overview the main points of gene expression analyses. We will illustrate using Affymetrix gene expression arrays and Illumina RNA-seq reads, but most of the underlying concepts port well to other platforms. Various preprocessing quality control metrics are discussed as well as how to evaluate the quality of the data. Next we discuss how to setup contrasts and detect differentially expressed genes.

5.1 Introduction to Gene Expression Analysis

A seminal paper in 1995 [97] introduced the concept of a massively parallel version of the Northern and Southern blot technique which allowed thousands of hybridizations to be conducted at the same time. This pioneering work revolutionized our ability to generate high quantities of system-wide, parallelized gene expression data. Instead of being constrained to studying gene expression of just a single gene or a handful of them at a time, researchers could now simultaneously study and quantify a large proportion of the RNA (primarily *mRNA*) expressed in a sample.

This opened the door to global expression profiling studies and enabled a wide range of comparative work to be conducted. The primary focus of these studies is to identify genes (or at least gene expression signals) associated with differences between traits (or *conditions*). For example, we can compare gene expression levels between normal and cancer tumor cells and try to identify which genes are differentially expressed between the two; we can evaluate expression levels of different tissues of an organism to understand which genes are expressed in which tissues; we can also study population variability, test how new drugs affect gene

---

Electronic supplementary material The online version of this chapter (doi: 10.1007/ 978-3-319-14475-7_5) contains supplementary material, which is available to authorized users.

© Springer International Publishing Switzerland 2015
C. Gondro, *Primer to Analysis of Genomic Data Using R*, Use R!,
DOI 10.1007/978-3-319-14475-7_5

expression levels, contrast resistant and susceptible individuals to identify genes that confer resistance (or susceptibility), test how exposure to environmental stressors affects gene expression levels, how gene expression changes over time (time-course analysis), and the list goes on. In livestock, global gene expression analysis has been used to search for genes involved in disease resistance, marbling, meat yield, feed intake, and many other traits. In a nutshell, most expression studies will, at least initially, try to identify the subset of genes that are *differentially expressed* between two or more conditions—simplistically, it is a contrast analysis. This will be the focus of this chapter—differential gene expression (*DGE*) analysis.

However, while the broad application of gene expression is to characterize which genes are expressed at what levels under a given condition, there is a wide range of other applications which include identification of novel SNP in coding regions which can later on be used in association studies; mapping of expression QTL (*eQTL*) which involves identification of genetic polymorphisms that change expression levels of RNA; use of expression data for prediction or diagnostic purposes (e.g., cancer classification); identification of co-expression networks (groups of genes that respond similarly to a stimulus) that can shed light on genes that previously had no known function or provide indications about new interactions in biological pathways.

5.1.1   Platforms for Expression Profiling

There are two main platforms for expression profiling: microarrays [99] and direct sequencing of transcripts (*RNA-seq*) [115]. Until a few years ago microarrays were the dominant platform but with the rapid pace of developments in next-generation sequencing methods, the latter rapidly replacing the microarrays.

Microarrays consist of a substrate (array) onto which thousands of probes are adhered to (oligonucleotides, cDNA, PCR products) and separated from each other by micrometric distances. Probes are spotted onto an array (can be a simple polylysine coated glass plate) from a microtiter plate using a robot. These are termed *spotted* microarrays. The target sample will then hybridize to the probe if they are complementary. Interpretation of the experiment is made possible by labeling the target with a fluorescent dye which can be detected and quantified quantitatively or semi-quantitatively.

Classically, spotted array experiments consist of a competitive hybridization between two samples with each one tagged with a different dye, generally the ubiquitous fluorophores cyanine-3 (Cy3) and cyanine-5 (Cy5). The slide is read by a scanner which uses lasers to excite the fluorophores (absorption of Cy3—550 nm and Cy5—649 nm) and, e.g., a CCD converts the analogue light signal of the fluorophores into a digital signal of light intensities (emission Cy3—570 nm and Cy5—670 nm) which are saved in *TIFF* graphical format. Since the hybridization is essentially competitive (the two target sets of RNA compete with each other to hybridize with the probes), the sample with a higher expression of

RNA will attach more to its probe and this is detected by the intensity of emission of the fluorophores. Scanned images can be viewed as an overlay of the two intensity files (green—Cy3, red—Cy5, each color is referred to as a channel) with the probes that did not hybridize in black (no expression in both samples), the probes in which the two samples have similar expression levels in yellow and variants from pure green (only the Cy3 labeled sample is expressed) to pure red (only Cy5). Note that the color scheme is really just a nod to the wavelengths in the light color spectrum.

A variant to the spotted arrays is the *GeneChip*, a commercial array developed by Affymetrix which uses a photolithographic process analogous to computer silicon chip production. Masks are used to control the light-driven synthesis of oligonucleotides (*oligos*) directly on the chip surface itself. Each oligo is up to 25 nucleotides long and there are up to 40 different oligos to detect a gene product on each chip. Around half the oligos are perfect matches to characteristic regions of a gene; the other half uses an oligo with a mismatch in position 13 which, at least in theory, allows detection of nonspecific hybridization and experimental noise. The difference between Affymetrix arrays and the spotted arrays as described above is that the former is not a competitive hybridization, i.e., a single sample is hybridized on an array, and intensity reads are absolute instead of relative (spotted arrays can also be used in this manner though, only Cy3 is used).

Spotted arrays are more noise prone than Affymetrix chips but they do allow for greater flexibility since any probe can be designed to spot on the array and individual research groups can build their own arrays *in house*—this is relevant for organisms that Affymetrix does not cover. Cost-wise spotted arrays are cheaper than *GeneChips* but a probe library is costly and time consuming to construct. There are also commercially available arrays from, e.g., Agilent for various species (and you can also order custom arrays) that are quite similar to conventional in-house spotted arrays but use long oligos (60-mer) and the printing technologies are much more reliable. Although this has a somewhat *historical feeling* to it, an overview of platforms and comparison of expression data using different platforms is given in [48] and [118].

RNA-seq uses next generation high-throughput sequencing platforms (e.g., Illumina) to sequence RNA transcripts and generates millions of short raw sequence reads which then have to be assembled (most commonly aligned against a reference); quantitated (number of reads that aligned to a genomic feature) and annotated (e.g., what gene do the transcripts belong to) [115]. RNA-seq holds similarity to an expression profiling method called *SAGE* (Serial Analysis of Gene Expression) which was already used with the early Sanger sequencing platforms to sequence and quantify RNA. While in the early days SAGE never really made inroads due to its prohibitive cost and time demands; the ever faster and cheaper sequencing platforms have enabled RNA-seq to become the platform of choice for expression studies. The basic difference between SAGE and RNA-seq is that the former sequences *tags* and not the whole transcript, only fragments of the transcripts are sequenced. A modern version of SAGE blends the use of short gene tags (26 bp) with high-throughput next-gen sequencers (HT-SuperSAGE) [72, 73].

In general terms, an RNA-seq experiment involves making a *library* prior to the sequencing. The basic steps will require RNA extraction, enrichment and fragmentation, followed by conversion to complementary DNA (*cDNA*, which is also what is used for microarray hybridization) and then platform specific proprietary adaptor sequences are attached to the ends of the fragments; finally an amplification round completes the library preparation step. *Barcodes* (short sequences 5–7 bp that are used to tag a library) are sometimes also added to allow multiplexing of samples (run multiple samples in the same sequencing reaction).

The key differences between microarrays and RNA-seq are:

- The costs of running an experiment with microarrays are still lower than RNA-seq on a per sample basis; but this may already not be the case anymore on a per feature comparison though. Tag reads instead of full sequencing reduces costs even further. Microarray costs have stagnated for quite some time while sequencing costs continue coming down at a fast rate; the cost balance will probably swap over in the very near future.
- Microarrays are much more mature in terms of platforms, analytical methods, computational tools and our understanding of possibilities/limitations. RNA-seq is more recent and almost the same questions we had with microarrays in the early 2000s are being asked all over again now with RNA-seq: what is the accuracy, how reliable is the data, what the error rates are and how to correct for them, what are the sources of bias, how to normalize the data, what are the appropriate statistical methods for the analysis.
- The dimensionality of raw RNA-seq data is many orders of magnitude larger than microarrays and demands high end computational resources and large storage availability.
- RNA-seq provides count data, the actual number of transcripts in a sample (not quite so simple since counts depend on sequencing depth, which is discussed in the next section). On the other hand, microarrays are essentially fluorescence intensities that are read by a scanner and constrained by its resolution— low expression signals cannot be resolved from background noise and high expression signals can *saturate* the scanner (i.e., everything above a certain level looks the same).
- Microarrays can only quantify expression levels of features that are on the array, anything else goes undetected. Microarrays also demand a reasonably solid prior knowledge of the organism of interest to select the probe sets. RNA-seq has the upper hand in that it is not constrained by preselected probe sets and any transcript that is expressed is potentially quantifiable. It is also easier to work with unconventional species for which there is no reference sequence available because RNA-seq provides the actual nucleotide composition of the RNA (as short read fragments though) which can be used in BLAST searches to try to find matches to other known organisms and obtain at least a rough annotation (not really so straightforward as this will usually require a rather tricky *de novo* assembly at some point). But still better than the anonymous probes used in early arrays that had to be characterized after the profiling experiment.

- RNA-seq can identify new transcripts, new variants such as SNP, alternative splicing events. Overall it is a broader and more flexible approach to explore gene expression.

So, are microarrays obsolete? Not yet. In 2008 Shendure [101] already forecast their diminishing importance in gene research but also envisioned a wider uptake in routine clinical applications, e.g., diagnostics (easier, cheaper and better established). Quite correctly, research purely focussed on global expression is steadily adopting RNA-seq; however, various studies in which global profiling is not the primary objective will continue to rely on microarrays as a research tool, at least for some time. A comparison between RNA-seq and microarrays is given in [122] and [124].

5.2   Experimental Design

A lot of effort has been put into optimization of experimental designs for microarrays. Most of this work focused on two color arrays since these are competitive hybridizations and it becomes important to set up the experiment in such a way as to maximize the power of the contrasts of interest (for an overview see [42]). With *one slide–one sample* arrays (single channel hybridizations) there are much less problems because instead of ratios we are actually working with intensities (there are ways of estimating channel intensities from two color array experiments but we will not discuss them herein) and it is much easier to add new data to an existing experiment (even though there are array batch specific effects that have to be accounted for).

If you must use two color arrays make sure that dye bias (there are differences between the intensities just due to dye effects) is not confounded with treatment, minimize the distance between contrasts—your primary question should be answered by hybridizations on the same slide and not across slides (the between slide variation is much larger than the within slide variation).

For RNA-seq the main concern is *depth of sequencing*. This refers to the number of reads that are sequenced. At a higher cost, a higher depth can be obtained and you will be able to identify rare variants and improve statistical power, or this can be reduced but there will be some loss. There is no exact formula for calculating depth and it is quite dependant on what the study objective is, but the discussions in [8, 50] and the R package *RNASeqPower* are good starting points. At the end of the day it is a trade-off between how many samples will be sequenced and at what depth for a given amount of money. Liu et al. [69] demonstrated that adding more depth after 10M reads does not greatly increase power to detect differentially expressed genes, on the other hand more biological replicates significantly increase power irrespective of the depth.

In terms of reproducibility both microarrays and RNA-seq are quite robust. There is not too much point in performing technical replicates of a sample. However, biological replicates are extremely important. Historically, expression studies are almost always underpowered. The data is noisy and the number of samples is small. While it is common practice with association studies to use thousands of samples, there are few expression studies with more than a hundred biological replicates (and these normally have ten different experimental treatments!). The problem here is that we are trying to do over 20,000 tests on, say 20 samples—a complete flip of the $p \times m$ matrix. The short version is: do as many biological replicates as your money will allow. The magic number is to aim for at least eight samples per treatment (remember the *statistics of small numbers*). Try to run all samples at the same time, same lab, same personnel, same equipment... Pay attention not to confuse pseudo-replicates (different samples from the same individual) and technical replicates with true biological replicates: there's not much point in using the same sample repeatedly unless you want to test the repeatability of the platform. Don't pool samples unless you have independent pools of the same treatment and you really do not have another option. In livestock (and all other projects for that matter) be careful with fixed (confounding) effects, for example if you are trying to find genes responsible for marbling in cattle don't contrast Brahman with Wagyu, all you will find out is which genes are differentially expressed between the two breeds—breed and marbling will be completely confounded. So, as a rule of thumb, try to use the most similar samples as possible that differ only in the trait of interest (but as much as possible for this trait if using small numbers).

5.3   Gene Expression Data

For the remainder of the chapter we will use Affymetrix microarrays and work through all the steps from raw data to final differential expression results. While not the most cutting edge technology, microarrays are still widely used and they are not overly demanding in terms of computational resources which make them more suitable for training purposes. I also believe that the principles that underpin array analysis provide a good entry point to RNA-seq analysis. Another point is that R cannot take raw sequence reads and run a full analysis; for RNA-seq some additional software is needed (mostly for mapping the reads to a reference). We will however discuss the general points of RNA-seq, provide pointers to relevant R packages and perform some basic tasks with a single small RNA-seq data file; at the end to illustrate differential expression with RNA-seq a small simulated dataset will be used. RNA-seq analysis will be *shadowing* the array analysis. A good overview of microarray analysis is given in [103]; the RNA-seq counterparts are [82] and [92]. A must read step-by-step tutorial style paper for analysis of RNA-seq using (mostly) R is [7].

5.4 Preprocessing and Quality Control

R is arguably the *de facto* platform for analysis of array data. There's a large number of packages available from Bioconductor and it's the strongest point of R in relation to analysis of genomic data. Here we are only working with Affymetrix arrays but the approaches we will discuss are similar for other array platforms, keep in mind however that each one has its own subtleties—before analyzing array data it is always a good idea to read the documentation of the platform you are working on, the array providers usually have extensive documentation available on their websites. Very nice worked out examples for other platforms (and also Affymetrix) are given in Hahne et al. [47] and Gentleman et al. [40]. There are many other books on analysis of microarrays (not necessarily using R), a nice read is Zhang's [123] book.

We will work with Affymetrix *GeneChip* bovine arrays. This expression array contains 24,128 probe sets representing 11,255 gene identities from Bos taurus build 4.0 and 10,775 annotated UniGene identities, plus 133 control probes. Each probe set consists of 25-mer short oligos with 11 *perfect matches* (*PM*) and 11 *mismatches* (*MM*), the base in position 13 is a mismatch. This set of oligos target the same gene but different *regions*. Our dataset consists of five control slides and five treatment slides (so much for my minimum of eight samples rule!). This is just a *toy* example so the control and treatment are not real, but the method is the same for any *contrast* you want to test.

Preprocessing and quality control are paramount in expression studies. In terms of quality control, the key objective is to identify slides of bad quality and remove them from further analyses. Preprocessing is another interesting feature of array studies which aims to remove technical *noise* from the data. With RNA-seq the objective is to remove reads (or parts of reads) of poor quality (reads that we think have sequencing errors) and sequences that do not belong to our samples (adapters and barcodes). We will discuss each of these, but first let's get our data into R.

5.4.1 Importing Gene Expression Data into R

R has no functions for reading intensities from scanned images, and this is best left to the manufacturers anyhow. Data from most platforms come as text files (the extension might be, e.g., *gpr*, but it's still just a text file). As such they can be imported into R using the usual *read.table* function. The problem is that many functions from the packages expect an *ExpressionSet*, this is just a container for your data with some rather nice features to hold experimental data together and extract/handle data, it's usually easy to convert a *data.frame* to an *ExpressionSet*, have a look at the library *convert* and implementation details of *ExpressionSet* in the help files for the library *Biobase*. Affymetrix data is a bit more complicated since the data files you'll need (*Cel*) are in binary format. Luckily the *affy* library has a function—*ReadAffy*—to read *Cel* files which makes small work out of the task (see also the *affxparser* package). Note that most of the packages we will use are from Bioconductor, remember to set the Bioconductor repositories (*select repositories*) or you will get an error saying that the package is not available and cannot be installed (these packages are not in *CRAN*). Once you have installed the *affy* package, to read our slide data all we have to do is

```
> library(affy)
> filenames=c(paste("ctrl",1:5,".CEL",sep=""),
+      paste("treat",1:5,".CEL",sep=""))
> Names= c(paste("C",1:5,sep=""),
+      paste("T",1:5,sep=""))
> slides=ReadAffy(filenames=paste("chapter5/",
+      filenames,sep=""),sampleNames=Names)
> print(slides)

AffyBatch object
size of arrays=732x732 features (19 kb)
cdf=Bovine (24128 affyids)
number of samples=10
number of genes=24128
annotation=bovine
notes=
```

That's it! We can see that we have ten arrays and they are bovine (*cdf=Bovine*). Before we go on we should mention the *cdf—chip definition file*. These are unique to Affymetrix, they describe the layout of the oligos on the slides and which ones form a probe set. There are some interesting discussions as to which *cdf* should be used; as annotations evolve over time, what used to be a good match for a certain gene might no longer be [37]. Various *alternative cdfs* have been developed (some of those are on Bioconductor) and you can even make your own (see the package *altcdfenvs*). Point is, whichever *cdf* you decide to use it will have to be in R's library path (if not, R will try to download the default *cdf* automatically). You can change the *cdf* using *slides@cdfName="nameofcdf"*.

There are many ways to access data in an *AffyBatch* object (see *affy* help files for more details). For example the perfect match intensities can be retrieved with *pm(slides)* and the mismatches with *mm(slides)*.

5.4.1.1   Importing RNA-Seq Data into R

With RNA-seq data it is usually easier to perform the QC, filtering and alignment steps out of R. A good starting point for the analysis in R is from BAM files. Bam files are the binary version of SAM (*sequence alignment/map*) files, a widely used plain text file format that contains information of aligned sequence data (see details in [68]). The R package *Rsamtools* provides functions to interface with *samtools* (a suite of utilities useful for post-processing of aligned reads) from within R.

We can however also read in raw sequence reads generated by the sequencer into R and do some initial processing of the files. The package *ShortRead* can be used to import *fastq* files into R and also has filtering and trimming functions [76]. Fastq files are text files with the sequence reads and quality scores for each nucleotide (plus additional information about the instrument, lane, multiplexing, pair read, etc.). Formats vary a little between platforms but the first line starts with @ and sequence information, second line are the raw sequence letters, third line is usually just a plus symbol (+), the last line has the quality scores for each nucleotide in the second line; the encoding varies, *Phred* or a scoring scheme based on it is common (check the documentation from the sequencer's manufacturer). It is important to know the correct scoring scheme so that the QC programs can correctly identify good/bad reads. An example with the first lines of a fastq file from an Illumina sequencer is shown in Fig. 5.1. In the book's folder for this chapter you will find a small example of a fastq file that we will use throughout the chapter for illustration purposes. Note that this is just a small part of a *real* fastq file; roughly, RNA-seq for one sample would hover around 2 Gb (much larger than the 5 Mb of our array data). Once you have installed *ShortRead*, the fastq file can be read into R with *readFastq*.

```
> library(ShortRead)
> seq=readFastq("chapter5/RNAseq.fastq")
> summary(seq)

Length       Class        Mode
102362  ShortReadQ        S4
```

Fig. 5.1 Example of a fastq file from an Illumina sequencer

The sequences can be visualized with

```
> head(sread(seq))

 A DNAStringSet instance of length 6
     width seq
 [1]     101 CCGCGAGCTACAGGCCCAGCTTCA...TCACAGCTGCCCTCA
 [2]     101 TTCAAGTTCTGACCCACTTCAAGG...ACTGCAGCCATGAGT
 [3]     101 GTCACATTCGAGTGGCGATACGGG...CATACGGGGACAAGG
 [4]     101 AAAACATGAATCTTAAAAAAAACG...CAGCTATCCTTCAAA
 [5]     101 GTTTGACAAAGGCTTTTGCCGGGC...TGTCTAGAGGGTTAC
 [6]     101 CAAAAATCCTTGATGACATCTTTG...CCCCAAGGACTGACC
```

A *real* fastq file might be too large to fit in memory, *ShortRead* has some functions that can help by breaking the file into chunks (*FastqStreamer*) or to sample subsets to check for problems (*FastqSampler*). To get only the sequence reads use *sread(seq)* and only the quality scores use *quality(seq)*. Check the documentation for additional details. Reads extracted using *sread* are of class *DNAStringSet* and can be manipulated using the *Biostrings* package (this package has a lot of useful functions to work with sequence data).

5.4.2   Quality Control

Once the data has been imported into R we can evaluate its quality. Microarrays are prone to exhibit high levels of experimental and systematic variability that are not related to the experimental contrasts. To ensure the best possible outcome it is critical that these effects are identified and adequately handled. Thus, the bulk of microarray analysis work lies in extensive preprocessing steps to determine the quality of the slides and calibration methods to remove spurious variation (the same as any other genomic dataset). Bad quality slides have unreliable intensity measures and can have a very large effect on final results. These slides should be identified and removed from the analysis. It is painful to exclude slides, each sample is expensive, some samples are hard to obtain and there usually are too few of them right from the start—but it just has to be done.

There are many ways to evaluate the quality of the data, we will see only a few here but the key principle is that the data should be reasonably uniform within and between slides—it's essentially a search for absence of patterns. Qualitative and quantitative quality control measures can be used to detect problematic slides, for example image plots for detection of spatial effects, relative and raw log expression intensities, slide correlations and others. Let's see how to do these in R.

A good way to generate image plots is with the package *affyPLM*, which fits a probe level model to the Affymetrix data. With *affyPLM* it is really easy to pick up spatial effects in the data which otherwise would be hard to notice.

```
> library(affyPLM)
> PLM=fitPLM(slides)
> par(mfrow=c(2,2))
> # image of log intensities
> image(slides[,1],main="log intensities")
> image(PLM, type="weights", which=1,
+      xlab=XLabel, main="weights")
> image(PLM, type="resids", which=1,
+      xlab=XLabel, main="residuals")
> image(PLM, type="sign.resids", which=1,
+      xlab=XLabel, main="sign of residuals")
```

First we fit a probe level model with *fitPLM* then we make four plots (Fig. 5.2) for the first slide. Initially we just plot the log intensities from the raw data, then the weights used in the regression to down-weigh outliers, then the residuals and finally just the sign of the residuals: one color for positive residuals and another

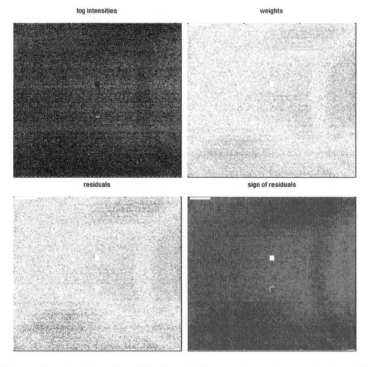

Fig. 5.2 Example of a bad quality slide. *Top-left*: log transformed raw probe intensities. The other three plots are derived from a fitted *PLM* model. *Top-right*: weights used to down-weigh outliers, *light green* shows areas of high weights and *dark green* low weights (outliers). *Bottom-left*: residuals. Negative residuals in *blue* and positive residuals in *red*. The intensity of the coloring reflects the value of the residuals, residuals close to zero are shown in *white*. *Bottom-right*: sign of residuals. Residuals with no intensity scaling—negative in *blue* and positive in *red*

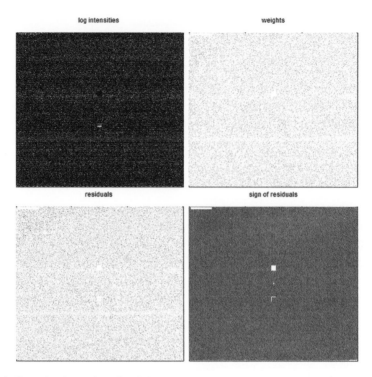

Fig. 5.3 Example of a good quality slide

color for negative residuals—makes it very easy to detect spatial effects. Note the large spatial effect in this slide, we already have a good candidate for a bad quality slide! Compare this slide with the slide (*ctrl5*) shown in Fig. 5.3—quite a difference! Notice also that it is much harder to pick up problems just from the raw log intensities plot.

Relative log expression is another commonly used QC measure (Fig. 5.4). Function *Mbox* will plot the *M* values for each array based on a pseudo-median array (details below). We would hope the medians are close to zero and the spread across arrays is similar.

```
> treatcol=c(3,3,3,3,3,2,2,2,2,2)
> Mbox(PLM,col=treatcol,
+      main="relative log expression",show.names=FALSE)
```

We used a variable *treatcol* to color code each treatment. Slides 1, 6, and 7 look somewhat worrisome.

We could also make boxplots or histograms of the raw log intensities (Fig. 5.5). Again we observe a fair bit of variability in the data set.

```
> par(mfrow=c(2,1))
> boxplot(slides, col=treatcol,
```

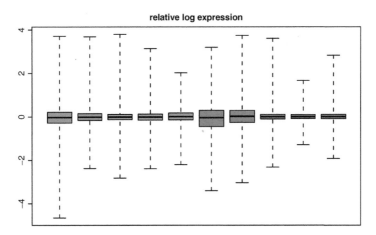

Fig. 5.4 Relative log expression plot from a fitted PLM

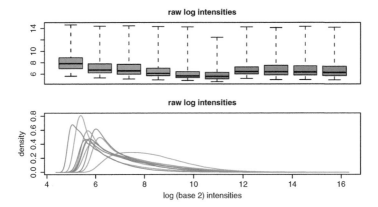

Fig. 5.5 Boxplot and histogram of raw probe level log intensities

```
+       main="raw log intensities", show.names=FALSE)
> hist(slides, col=treatcol,
+       lty=1,xlab="log (base 2) intensities",
+       main="raw log intensities")
```

It is also useful to plot the correlations between slides (Fig. 5.6). This can help identify outlier arrays and detect if the sample treatments group together—in principle slides within the treatment group should be more similar than across treatments (but not necessarily always, depends on the differences between the groups). To exemplify we use the perfect match intensities and *matrixPlot* from the library *ABarray* for graphing.

```
> library(ABarray)
> Cor=pm(slides)
```

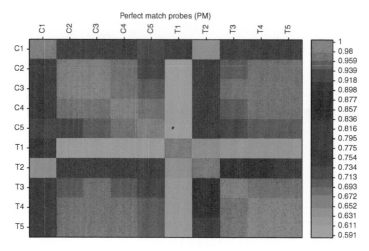

Fig. 5.6 Plot of array–array Pearson perfect match (PM) probes intensity correlation coefficients

```
> Cor=cor(Cor)
> matrixPlot(Cor, nrgcols=21, k=21,
+        title="Perfect match probes (PM)")
```

There are many other metrics, e.g., principal component analysis, RNA degradation plots, ratio of PM and MM expression intensity, MA plots (we'll discuss these later), check control probes. In spotted arrays we can look at signal-to-noise ratios, block effects, print-tip effects, dye effects, etc. See [47] and [40] for worked out examples.

Our data is definitely nothing to write home about. There are spatial effects, lots of variation between slides, but that's just the nature of real data. I would probably exclude slides one and six (and look more closely at seven) from further analysis or at least repeat the analysis with and without them and compare the results, if they are the same good, if not you will have to decide if the difference is due to the slides being bad or because you removed 20 % of your data—a tough call! As a rule of thumb, even though there are no well-defined criteria as to when an array should be discarded/redone, the combined QC measures can provide sufficient diagnostics for an empirical decision, ergo if a slide looks bad across several different QC tests, get rid of it.

5.4.2.1 Quality Control of RNA-Seq

With the RNA-seq data, a popular non-R program for quality control is *fastQC*. It is a Java program that, for each fastq file, generates a comprehensive html report with various metrics; e.g., quality scores per position and per sequence, GC content, adapter contamination, among others. As a reference, you will find the fastQC report for this data in the chapter's folder. FastQC can read fastq files as well as BAM and SAM. While highly informative, these reports can become unwieldy in large projects with hundreds or thousands of fastq files (one report for each file). The program will run on the command line, but it also has a nice graphical interface which makes it quite easy to use. FastQC is freely available from

http://www.bioinformatics.babraham.ac.uk/projects/fastqc/

Note that fastQC is a Java program and you will need the Java runtime installed on you machine for it to run. You can download Java from

https://java.com/en/

A similar job can be done in R with *ShortRead* with the *qa* function. By default only one million reads from each file will be used to create QC summaries. The returned object is of class *FastQA* and contains information on number of reads, base calls, quality scores, among others. To run QC on our data

```
> seqQC=qa("chapter5/RNAseq.fastq")
> seqQC

class: FastqQA(10)
QA elements (access with qa[["elt"]]):
  readCounts: data.frame(1 3)
  baseCalls: data.frame(1 5)
  readQualityScore: data.frame(512 4)
  baseQuality: data.frame(94 3)
  alignQuality: data.frame(1 3)
  frequentSequences: data.frame(50 4)
  sequenceDistribution: data.frame(57 4)
  perCycle: list(2)
    baseCall: data.frame(426 4)
    quality: data.frame(3618 5)
  perTile: list(2)
    readCounts: data.frame(0 4)
    medianReadQualityScore: data.frame(0 4)
  adapterContamination: data.frame(1 1)
```

The various elements from the *FastqQA* object can be retrieved with, e.g.

```
> seqQC[["readCounts"]]

                 read filter aligned
RNAseq.fastq 153557     NA      NA

> seqQC[["baseCalls"]]

                   A       C       G       T   N
RNAseq.fastq 4169015 3619064 3585408 4135305 465
```

QC results can be used for additional downstream summaries that you create yourself. Or you can use the *report* function from *ShortRead* to create an html report and save it in the directory of Chap. 5

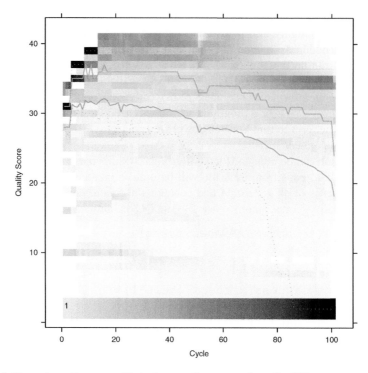

Fig. 5.7 Per cycle quality scores. Notice how quality starts to drop after 80 bp

```
> report(seqQC,dest="chapter5/QCreport")
```

```
[1] "chapter5/QCreport/index.html"
```

If you open the report you will see that it is based on a template and the comments do not necessarily generalize across platforms. It is not too hard to change the template for your own purposes/platform though. A good place to begin evaluation of the data is to check for adapter contamination (need to exclude, will cause alignment problems). Next see if the read quality is acceptable across all *cycles* (think in terms of nucleotide position *1–n*); at the beginning and end of reads the quality tends to drop down—you might need to trim a few bases on each side (again a tradeoff, reads become shorter and harder to align but of better quality or longer but unreliable). Figure 5.7 shows the quality score per cycle, notice how quality starts to drop after around 80 bp—in this case you might want to trim the last 10–20 bases. Also check quality of the sequences themselves (how many have high quality scores). These are the basics, but there are many other metrics you could evaluate, e.g., the distribution of nucleotide composition and GC content to check if they are within what you would expect.

A bit cumbersome and time demanding but worthwhile doing is to run QC on the raw sequences, filter the data based on parameters that you think are reasonable given the quality of the data and then run the QC all over again to see if the data now looks better.

5.4.3 Preprocessing

Back to the arrays. Slides deemed of adequate quality will then undergo calibration steps which generally consist of: (1) background correction to remove intensity measures that are not due to the target; (2) normalization which is necessary for across array comparisons, achieved by adjusting the overall distribution intensities making them similar across slides (note that this step usually makes a dataset testable only within itself, if new slides are added to the experiment, the entire set has to be renormalized); and (3) a summarization step which is more specific to Affymetrix GeneChips since these are unique in the use a set of short oligos to target a transcript. This probe set has to be summarized into a single intensity value for each target on each array. Different methods have been developed for each of these steps [59] and, even if it is still unclear which approach is best, it has been shown that the main source of variation between results is due to the choice of summarization method [49]. It's probably a good idea to run the analysis with different summarization methods and compare the results.

Various summarization algorithms have been proposed. The main ones are: MAS 5.0 [2], RMA [57, 58], GCRMA [121], PLIER [3], VSN [56], and MBEI [67]. We will not discuss the algorithms here, just focus on how to use them in R. If you have to pick a favorite, RMA is not a bad choice, it's statistically sensible and has been widely used in the most diverse range of projects.

There are direct wrappers to most of the summarization methods as shown below. For MAS 5.0, RMA, and GCRMA you'll need the library *affy*, for plier the package *plier*, for VSN *vsn* and for MBEI you can use *expresso* in the *affy* library. *mas5*, *rma* and *gcrma* are wrappers for *expresso*, these methods could be called using *expresso* with a *summary.method* argument. Check the documentation for the various parameters and possible combinations for each summarization method, as for example the summarization call to MBEI shown below (e.g., background correction method, normalization method...).

```
> library(affy)
> MAS=mas5(slides,sc=200)

background correction: mas
PM/MM correction : mas
expression values: mas
background correcting...done.
24128 ids to be processed
|                   |
|###################|

> MAS =exprs(MAS)
> MAS =log2(MAS)
> MASCalls=mas5calls(slides)
```

```
Getting probe level data...
Computing p-values
Making P/M/A Calls

> MASCalls =exprs(MASCalls)
> RMA=rma(slides)

Background correcting
Normalizing
Calculating Expression

> RMA =exprs(RMA)
> GCRMA=gcrma(slides)

Adjusting for optical effect.....Done.
Computing affinities.Done.
Adjusting for non-specific binding.....Done.
Normalizing
Calculating Expression

> GCRMA=exprs(GCRMA)
> library(plier)
> PLIER=justPlier(slides,normalize=TRUE)

Quantile normalizing...Done.

> PLIER=exprs(PLIER)
> library(vsn)
> VSN=vsnrma(slides)

vsn2: 535824 x 10 matrix (1 stratum).
Please use 'meanSdPlot' to verify the fit.

Calculating Expression

> VSN=exprs(VSN)
> MBEI = expresso(slides,
+       normalize.method="invariantset",
+       bg.correct=FALSE,
+       pmcorrect.method="pmonly",
+       summary.method="liwong")

normalization: invariantset
PM/MM correction : pmonly
expression values: liwong
normalizing...done.
24128 ids to be processed
|                        |
|####################|
```

```
> MBEI=exprs(MBEI)
> MBEI=log2(MBEI)
```

So, what did we do? We summarized the data using each of the methods. Then we used the function *exprs* to extract the intensity values from the *ExpressionSet* object as a matrix of *probes* × *samples*. For example

```
> print(MAS[1:5,1:3])
```

```
                     C1        C2        C3
AFFX-BioB-3_at 7.382151 7.968781 8.026792
AFFX-BioB-5_at 8.641223 7.735663 8.224185
AFFX-BioB-M_at 8.721451 8.333241 8.744490
AFFX-BioC-3_at 9.656310 7.623355 8.421468
AFFX-BioC-5_at 9.278536 7.800786 8.248430
```

Notice that sometimes we converted the expression intensities into *log2* and at other times we did not. RMA, GCRMA, PLIER, and VSN already return intensities in log 2 scale (usually an adequate data transformation for array data—it is a natural fit to digital imaging data which is coded in bits).

One last thing to notice is the function *mas5calls*. This returns a matrix of the same dimension of MAS with flag calls for the expression of each probe in each sample. The flags are *P—present, M—marginal,* and *A—absent.* These are handy to filter the data before testing for differential expression, for example, remove all probes that are flagged as marginal or absent in all arrays. This reduces the number of tests to be carried out—less problems with multiple testing; and further, low intensity probes tend to be unreliable since they are at the border of detection of the scanner, it's good to remove them.

```
> print(MASCalls[1:5,1:3])
```

```
                C1  C2  C3
AFFX-BioB-3_at "M" "P" "P"
AFFX-BioB-5_at "P" "P" "P"
AFFX-BioB-M_at "P" "P" "P"
AFFX-BioC-3_at "P" "P" "P"
AFFX-BioC-5_at "P" "P" "P"
```

Before we test our datasets for differential expression we can check what the normalized and summarized data looks like. We will use RMA to exemplify.

```
> par(mfrow=c(2,1))
> boxplot(RMA, col=treatcol,
+     main="Boxplot of normalized log intensities",
+     show.names=FALSE)
> plot(density(RMA[,1]),col=treatcol[1],
+     main="Histogram of normalized log intensities ",
+     xlim=c(min(RMA),max(RMA)),xlab="log2 intensity")
```

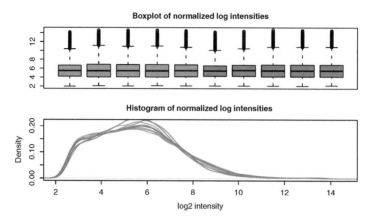

Fig. 5.8 Boxplot and histogram of normalized probe set log intensities

```
> for (j in 2:length(treatcol))
+     lines(density(RMA[,j]),col=treatcol[j])
```

If we compare the normalized data in Fig. 5.8 with the data in Fig. 5.5 we see
that the normalization steps really did their job—we've stabilized the means and
the variances. Just be careful because the normalization normally works but that is
not an indication that your slides are ok now, there's only so much data bending
you can do. Another interesting graph is to plot the first two principal components
of the data (Fig. 5.9), we would expect that the samples from the same treatment
would be more similar to each other and cluster together while the distances between
treatments would be greater—this is not entirely the case here (but not really very
bad either). This is another indication that our data is not of very good quality or that
the contrast is not very large. Be careful with the interpretation of the PCA, if there
are a lot of differentially expressed genes the treatments will separate well (and it
is an early indication of this), otherwise not. If you notice that the samples group
into, e.g., dates of running the hybridization or slide batches (could fit these effects
as a color code), then you have a concern. Slides that are very far apart from others
(irrespective of treatment) also should be looked at more closely.

```
> PCA=princomp(RMA)
> PCA=loadings(PCA)[,1:2]
> plot(PCA,col=treatcol,main="Principal components plot",
+     pch=treatcol,xlab="PCA1",ylab="PCA2")
```

In Fig. 5.9 we see that three slides are quite different from the others. We can find
these slides quite easily using *which*

```
> which(PCA[,1]> -0.31)
```

```
C1 T1 T2
 1  6  7
```

And the same three slides we had flagged before show up again…

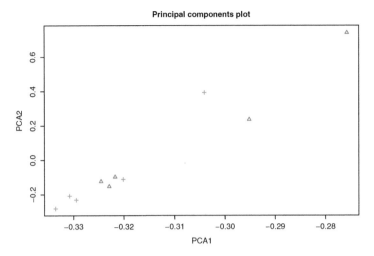

Fig. 5.9 Plot of first and second principal components for RMA normalized data

5.4.3.1   Preprocessing of RNA-Seq

We left off the RNA-seq data after running QC with *ShortRead* and we noticed that the quality was not too good after 80 cycles. The next step is to filter the data by removing adapters, trim low quality cycles, and remove low quality reads. Again this is generally easier to do out of R. A good program for our Illumina data is *trimmomatic* [12]. The program is available from

http://www.usadellab.org/cms/?page=trimmomatic

Conveniently it includes the adapters that Illumina uses, making life much easier. Trimmomatic also requires the Java runtime. Copy trimmomatic and the adapters to your working directory, from the command prompt (*not* from R) go to the working directory (using, e.g., *cd c:\primer*) and then run

```
Java -jar trimmomatic-0.32.jar SE -phred33
    chapter5\RNAseq.fastq chapter5\trimmedRNA.fastq
ILLUMINACLIP:adapters\TruSeq3-PE.fa:2:30:10
LEADING:3
TRAILING:3
SLIDINGWINDOW:4:15
MINLEN:50
HEADCROP:10
CROP:80
```

Note that the command has to be on a single line, but here for clarity it is split into multiple lines. This will run trimmomatic with our RNA-seq file and output filtered results into a new fastq file called *trimmedRNA.fastq*. There are quite a few parameters here (for full details check the help files of the program) but briefly:

SE means single end reads, for paired-end data use *PE* (note: this example is from paired end data but we are using just one file so let us pretend it is single end). Next is the quality scoring scheme (*phred32* or *phred64*), then the names of the input and output files following by trimming parameters. *ILLUMINACLIP* points to the file with the adapters; *LEADING* and *TRAILING* remove bases at the beginning and end of the reads below a score of 3; *SLIDINGWINDOW* slides across the sequence and removes those inside a window of length 4 that have an average quality below 15; *MINLEN* removes sequences that are shorter than 50 after filtering; *HEADCROP* removes the first 10 sequences at the start of the run (check the QC report—there is a lot of nucleotide content variation at the start of the sequences) and finally *CROP* deletes everything after the first 80 bases (recall the quality was not so good after 80 cycles). Note that commands are sequential. This leaves us with a filtered fastq file with 112,364 reads (73.17%) of the initial reads and you will notice that the length is shorter due to edge trimming and sizes vary (originally they were all the same length). If you run a QC report again (the fastQC report is in the folder) you will see that the metrics are much better now, but it was a hard pruning—probably excessive. Run QC with *ShortRead* to compare the results.

```
> seqQC=qa("chapter5/trimmedRNA.fastq")
> report(seqQC,dest="chapter5/trimmedQCreport")

[1] "chapter5/trimmedQCreport/index.html"
```

Short Read can also be used for data filtering. It has some in-built functions and it also allows users to create their own. Check the documentation for details, but just to illustrate

```
> seq=readFastq("chapter5/RNAseq.fastq")
> seq

class: ShortReadQ
length: 153557 reads; width: 101 cycles

> seqF=seq[nFilter()(seq2)]
> seqF=trimEnds(seqF,"I")
> seqF

class: ShortReadQ
length: 153557 reads; width: 0..80 cycles
```

We still need to align the data against a reference genome. This has to be done out of R and a good option is *bowtie2* [66]. For this you will need bowtie2 installed and an index for reference genome (which is quite large, in excess of 3 Gb). You can make your own index from raw sequence data and use bowtie2 to build an index or you can download one that is already available. We will not go into details of bowtie2 but the software and some indexes can be downloaded from

http://bowtie-bio.sourceforge.net/bowtie2/index.shtml

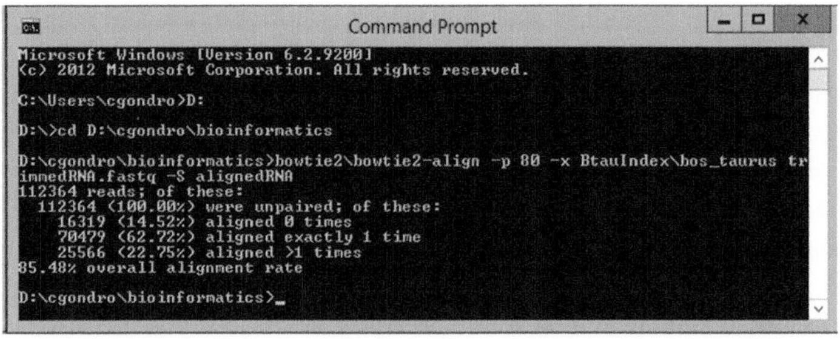

Fig. 5.10 Snapshot of the console using bowtie2 to align the trimmed RNA-seq data against the reference genome

Additional indexes can be downloaded from Illumina:

http://support.illumina.com/sequencing/sequencing_software/igenome.html

The commands to align our data and the output from bowtie2 are shown in Fig. 5.10. The aligned data is in SAM format.

The last step is to convert the SAM file into BAM, sort and index it (makes searches much faster and is a requirement of some programs). For this we use *samtools* [68] which can be downloaded from

http://www.htslib.org/

Install samtools in the working directory and from the command line (console, not R) run

```
samtools\samtools.exe import
reference\bos_taurus.fa.fai
chapter5\alignedRNA.sam chapter5\alignedRNA.bam

samtools\samtools.exe sort
chapter5\alignedRNA.bam chapter5\sortedAlignedRNA

samtools\samtools.exe index
chapter5\sortedAlignedRNA.bam
```

Again all three are single line commands, just split for convenience. You should end up with a final dataset ready to go (*sortedAlignedRNA.bam* and the index has the same name with extension *.bai*). Note that you will again need a reference genome to convert from SAM to BAM. An additional step might involve merging files to concatenate data from a sample that is spread across two or more SAM/BAM files. Use the *merge* command from samtools.

5.5   Analysis of Differential Expression

Returning to the arrays, we start by filtering out the control probes and the probes that were flagged as *absent* or *marginal* in over half of the samples. The control probes are what they sound—controls. They are extraneous to the target organism and at least in principle they should not be differentially expressed between treatments (they are also spiked-in at different concentrations and can be used to test intensities). The truth is, more often than not these controls will pop up as differentially expressed, so once you are through with them in the QC stages remove them. The *A/M* flagged probes should be removed to reduce the number of tests. Give some consideration to the cut off threshold. In our example, if a probe is expressed in one treatment but not in the other it would be reasonable to expect five slides flagged as *A/M* and these are probably the probes you are most interested in—expressed in one treatment, not expressed in the other. Either separate the calls per treatment and then decide on a cut off or at least make sure that you do not penalize non expressed treatments. I'm stressing this point because the *P/M/A* flags are commonly mistaken with bad quality reads and they should not be interpreted as such. There's a difference between 50 % are bad and 50 % are not expressed. There are other options for filtering the data. Instead of flag calls use a filtering criterion based on, e.g., variance, expression intensity, etc. The *genefilter* library has some nice functions, particularly *nsFilter*. Other filtering options depend on what you intend to do with the data downstream, if you want to associate probes to genes/function you will need the probes to be annotated (see Chap. 6), if there are unannotated probes on the array and they are of no interest they can be filtered out as well (duplicates, different probes that target the same gene can also be filtered out—but I tend to keep them, at least for differential expression testing).

We will continue using the RMA normalized data, but it's exactly the same for all other datasets. The filtering steps are

```
> dim(RMA)

[1]  24128     10

> index1=grep("AFFX",row.names(RMA),ignore.case=TRUE)
> length(index1)

[1]  133

> Pcounts=apply(MASCalls,1,
+      function(x) length(which(x=="P")))
> index2=which(Pcounts<=4)
> length(index2)

[1]  15123

> summary(factor(Pcounts))
```

0	1	2	3	4	5	6	7	8
10396	2017	1176	804	730	650	688	816	1115

9	10
1435	4301

```
> fRMA=RMA[-unique(c(index1,index2)),]
> dim(fRMA)
```

```
[1] 8945    10
```

First we use *grep* to identify the control probes (Affymetrix uses *AFFY* in the probe names), then we use *apply* to count the number of *P* calls for each probe and get the indices of those with 4 or less *Ps* per probe (6 will be *A* or *M*). We see that a large number of probes are absent in all samples. After filtering we have only *8,945* probes left.

Now we are ready to test for differential expression (DE). There are many ways of testing for DE ranging from simple mean fold change differences between treatments to mixed models with various effects (slides, probes...). We will use linear model modeling from the package *limma* [104]. *limma* has been widely used in array studies and is a very mature and easy to use package. There is an excellent online manual which covers not only Affymetrix arrays but also analysis of spotted arrays and data preprocessing. To use *limma* with Affymetrix arrays is straightforward, all we need is our data, a design matrix, and a contrasts matrix to define the comparisons of interest. The design matrix simply states which arrays belong to which treatment, the contrasts matrix define which treatment combinations we want to test. In our case the design matrix is

```
> Design=cbind(c(rep(1,5),rep(0,5)),
+        c(rep(0,5),rep(1,5)))
> colnames(Design)=c("ctrl","treat")
> print(Design)
```

```
        ctrl treat
 [1,]     1     0
 [2,]     1     0
 [3,]     1     0
 [4,]     1     0
 [5,]     1     0
 [6,]     0     1
 [7,]     0     1
 [8,]     0     1
 [9,]     0     1
[10,]     0     1
```

And the contrasts matrix is just as simple

```
> Contrasts=matrix(c(1,-1),byrow=F)
> colnames(Contrasts)="ctrl-treat"
> rownames(Contrasts)=c("ctrl","treat")
> print(Contrasts)
```

```
      ctrl-treat
ctrl           1
treat         -1
```

All we did was define which samples belong to each treatment in the design matrix and in the contrasts matrix we defined that we want to test *control* × *treatment* (not as if we had anything else to test anyhow). The order of the terms is important to interpret which probes are over or under expressed. I'm not fond of the terms up regulated and down regulated, that depends on how the contrast is set up. For a *classic control* × *treatment* it might make sense (remember that we would have to swap the contrasts) but for anything else it can be a bit confusing. Using the order of the terms makes it easy to interpret results in any way which is convenient. For example in our case we know that the contrast is *control–treatment*, hence any positive difference in expression means that the control is more expressed than the treatment and any negative difference means the treatment is more expressed than the control. Note: here we will use the full *RMA* dataset, but we should of course use the filtered subset (*fRMA*); this is just to illustrate high and low expression differences during the analysis (as an exercise repeat all the next steps later on with *fRMA* instead). To test for differential expression we use

```
> library(limma)
> Fit=lmFit(RMA,Design)
> Fitc=contrasts.fit(Fit,Contrasts)
> Fitb=eBayes(Fitc)
```

lmfit is used to model the data and is quite similar to *lm* as we have used before but results are stored in a *MarrayLM* object which is a bit different from the *lm* object. But we can still retrieve the coefficients from *Fit* using

```
> head(Fit$coefficients)
```

	ctrl	treat
AFFX-BioB-3_at	7.838531	7.711012
AFFX-BioB-5_at	7.514108	7.254552
AFFX-BioB-M_at	8.122386	8.002476
AFFX-BioC-3_at	8.438084	8.204015
AFFX-BioC-5_at	7.832757	7.709140
AFFX-BioDn-3_at	11.314343	11.424412

Next we tested our contrast (which here is just the difference *ctrl–treat*). Again we can retrieve the coefficients with

```
> head(Fitc$coefficients)
```

	ctrl-treat
AFFX-BioB-3_at	0.1275193
AFFX-BioB-5_at	0.2595556
AFFX-BioB-M_at	0.1199100
AFFX-BioC-3_at	0.2340691
AFFX-BioC-5_at	0.1236175
AFFX-BioDn-3_at	-0.1100690

And to test for differential expression we used the function *eBayes* which uses an empirical Bayes shrinkage of standard errors to calculate the statistics [104]. We can retrieve the results from the moderated *t*-test and corresponding *p*-values using

```
> head(Fitb$t)

                 ctrl-treat
AFFX-BioB-3_at    0.3595526
AFFX-BioB-5_at    0.7051612
AFFX-BioB-M_at    0.3181206
AFFX-BioC-3_at    0.6716419
AFFX-BioC-5_at    0.3544924
AFFX-BioDn-3_at  -0.3922216

> head(Fitb$p.value)

                 ctrl-treat
AFFX-BioB-3_at    0.7249713
AFFX-BioB-5_at    0.4931869
AFFX-BioB-M_at    0.7554603
AFFX-BioC-3_at    0.5136039
AFFX-BioC-5_at    0.7286702
AFFX-BioDn-3_at   0.7012672
```

If there were more contrasts we could also retrieve the *F*-statistics and their *p*-values with (*Fitb$F* and *Fitb$F.p.value*). Here of course there's a single contrast so both are the same. Additional arguments can be passed to *eBayes* such as the prior for the proportion of probes that we suppose will be DE, upper and lower limits of fold change standard deviations and the minimum fold change that we think is significant in the experiment. There are other slots of interest in an *MArrayLM* object (see the help files for details), but one that we should mention is *Amean*.

```
> head(Fitb$Amean)

 AFFX-BioB-3_at   AFFX-BioB-5_at   AFFX-BioB-M_at
       7.774772         7.384330         8.062431
 AFFX-BioC-3_at   AFFX-BioC-5_at  AFFX-BioDn-3_at
       8.321050         7.770949        11.369377

> summary(Fitb$Amean)

  Min. 1st Qu.  Median    Mean 3rd Qu.    Max.
 2.278   4.167   5.500   5.612   6.763  14.440
```

This is the mean expression intensity across all arrays for each probe. This is one metric you'll want to keep an eye on. Signals of low intensity tend to be unreliable—if a DE probe is in the low signal region it should be treated with some caution. Don't forget we are working on a log2 scale here—anything below 8 is not really a strong signal, it is only 256 on a linear scale; the upper bound on a 16 bit scanner (this

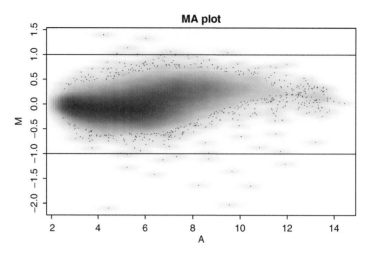

Fig. 5.11 MA plot

example) is 65,536 and of course, the maximum intensity value we can observe is 16 on a log2 scale. The mean intensity is usually referred to as *A* in array studies. Alongside *A* we can calculate the fold change difference between the two treatments, this is referred to as *M*. If we plot one against the other we have an *MA* plot, which as we mentioned above can also be used for QC purposes. Hence an *MA* plot such as the one in Fig. 5.11 can be made with

```
> smoothScatter(Fitb$Amean,Fitb$coefficients,
+      nrpoints=500,xlab="A",ylab="M",cex.main=0.9,
+ main="MA plot")
> abline(h=1)
> abline(h=-1)
```

The function *smoothScatter* is used just because it makes a prettier graph than *plot*. We added two horizontal lines to indicate fold changes above (or below) which reside the probes of interest (note that a rather low fold change was used). Another common plot is a *volcano plot* (Fig. 5.12), which is the minus log odds (LOD) of the *p*-values by the fold change (*M*) values.

```
> lod=-log10(Fitb$p.value)
> o1=which(Fitb$coefficients>1 | Fitb$coefficients< -1)
> o2=which(Fitb$p.value<0.01)
> o=intersect(o1,o2)
> plot(Fitb$coefficients, lod,
+      xlab="difference in expression (M)",
+      ylab="t-statistic p-values (-log10)",
+      cex.main=0.95, cex=0.25, main="volcano plot")
> points(Fitb$coefficients[o1],lod[o1],
```

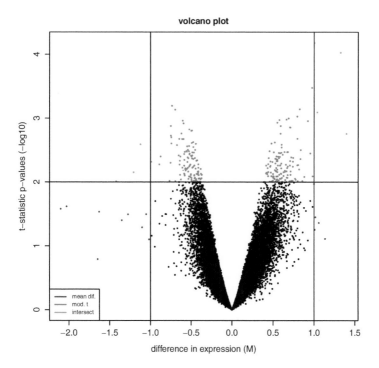

Fig. 5.12 Volcano plot

```
+        pch=19,cex=0.25,col="blue")
> points(Fitb$coefficients[o2],lod[o2],
+        pch=19,cex=0.25,col="red")
> points(Fitb$coefficients[o],lod[o],
+        pch=19,cex=0.25,col="green")
> legend("bottomleft",c("mean dif.",
  "mod.t","intersect"),
+        lty=1,col=c("blue","red","green"),cex=0.7)
> abline(h=-log10(0.01))
> abline(v=1)
> abline(v=-1)
```

It looks complicated but that's only because I wanted some nice colors in the plot. What the volcano plot shows us is a grid of *p*-values by fold changes and we can focus on the probes that are statistically significant and at the same time meet our fold change requirements (the green points in Fig. 5.12).

Basically we are finished. But, which after all are the differentially expressed probes?

```
> res=data.frame(FoldChange=Fitb$coefficients,
+      p.value=Fitb$p.value, Amean=Fitb$Amean)
```

```
> names(res)=c("FoldChange", "p.value", "Avalue")
> res=res[order(res$p.value),]
> head(res)

                  FoldChange        p.value   Avalue
Bt.3289.1.S1_at    1.0083137  6.613378e-05  5.682473
Bt.28515.1.A1_at   1.3295861  9.389443e-05  6.022375
Bt.22867.2.A1_at   0.9768392  3.347976e-04  7.797295
Bt.132.1.S1_at    -0.7326319  6.397408e-04  4.782339
Bt.25089.2.A1_at   0.8262308  7.246055e-04  6.324426
Bt.28987.1.S1_at  -0.6930765  7.362446e-04  4.644442

> length(which(res$p.value<0.01))

[1] 246
```

There, we made a data.frame with the probe names, fold changes, *p*-values, and A values. Then we sorted the results by *p*-value and selected a *p*-value cut off of *0.01*. Different filters can of course be applied—higher or lower *p*-values, *p*-values plus a minimum fold change, nothing below an A value of 8, and so on...

```
> length(which(res$p.value<0.01 & res$Avalue>8))

[1] 38
```

Most of our DE probes are in the low signal region—not a good sign.

5.5.1   Multiple Testing

Multiple testing is also a problem with microarrays. We discussed these issues in Chap. 3. To keep it simple, Bonferroni correction is probably too stringent for most cases, a reasonable compromise option is the Benjamin and Hochberg method which controls false discovery rates.

```
> adjusted=p.adjust(res$p.value, method="BH")
> length(which(adjusted<0.05))

[1] 0
```

Not a single probe survived multiple testing correction. Given the bad quality of three slides and the fact that we did not have a real contrast, it's probably a good thing! Some level of judgment comes into play with multiple testing, you might want to call it quits at this point and move on to the next project or you could go back to the original list or at least those 38 probes that had a reasonable signal and investigate if they have some functional significance that is worthwhile pursuing further. Now would be a good time to return to the beginning, exclude the three bad arrays, and redo the DE analysis with the filtered RMA data (*fRMA*). Functional work is the topic of the next chapter.

5.5.2 Differential Expression of RNA-Seq

Back to our RNA-seq data. Again we will just go over some basic pointers to get started with the analysis (we also only have one sample—so, nothing to test!). We still do not have the actual counts of transcripts, the first step is to summarize the BAM alignments into a count of features. For this we need to install some additional Bioconductor packages that can read the BAM file (*GenomicAlignments*) and create a database of transcripts (*GenomicFeatures*) that will allow matching sequence reads to functional annotations. The database is built on the fly by downloading (need to be connected to the internet) publicly available data (UCSC in this example):

```
> library(GenomicFeatures)
> library(GenomicAlignments)

> txdb=makeTranscriptDbFromUCSC(genome='bosTau6',
+         tablename='ensGene')

Download the ensGene table ... OK
Extract the 'transcripts' data frame ... OK
Extract the 'splicings' data frame ... OK
Download and preprocess 'chrominfo' data frame ... OK
Prepare the 'metadata' data frame ... OK
Make the TranscriptDb object ... OK

> txGene=transcriptsBy(txdb,'gene')
```

The various public databases (e.g., *ensembl* from EMBL-EBI) store information on genomic coordinates from reference genomes and associations with various forms of annotation information. Directly from R we can download this data and build an object to store our annotation needs. The *GenomicFeatures* package has a lot of functions and methods for assembly and manipulation of transcript centered annotation data. We used *makeTranscriptDbFromUCSC* to make a transcript database for cattle (*txdb*) and then grouped the features by *gene* identifiers (*txGene*). The BAM file holds information about the position of our RNA-seq reads in the cattle reference; and now we have an R object that can match these positions to functional information. Have a look at the contents of *txGene*, it holds the annotation and mapping information to match genes with locations from the BAM file. And to retrieve the actual counts of transcripts

```
> reads=readGAlignmentsFromBam(
+    "chapter5/sortedAlignedRNA.bam",sep="")
> newnames=paste("chr",rname(reads),sep="")
> newnames=gsub("MT","M",newnames)
> reads=GRanges(seqnames=newnames,
+    ranges=IRanges(start=start(reads),
+    end=end(reads)),
```

```
+            strand=rep("*",length(reads)))
> counts=countOverlaps(txGene,reads)
```

First we used *readGAlignmentsFromBam* (in *GenomicAlignments*) to read in the BAM file. The other two lines are a bit of a workaround to match names in the BAM file with names in *txGene*. If you looked at *txGene* you will have noticed that the chromosome names are, e.g., *chr1, chr2,...* But names in the BAM file (*reads*, below) are just *1, 2,...* We had to add *chr* to the names so that searches match (same with mitochondria—*MT* and *M*).

```
> reads
```

```
GAlignments with 96045 alignments and 0 metadata columns:
          seqnames strand           cigar      qwidth          start
             <Rle>  <Rle> <character> <integer>      <integer>
      [1]        10      -        70M5I5M          80         427292
      [2]        10      +            80M          80         669284
      [3]        10      -            80M          80         925717
      [4]        10      +            80M          80         944619
      [5]        10      +            41M          41        1072654
      ...       ...    ...            ...         ...            ...
 [96041]         X      +            80M          80      147944551
 [96042]         X      -            80M          80      147944759
 [96043]         X      +            80M          80      148270052
 [96044]         X      -            80M          80      148796047
 [96045]         X      +            80M          80      148808671

              end       width       njunc
        <integer>   <integer>   <integer>
           427366          75           0
           669363          80           0
           925796          80           0
           944698          80           0
          1072694          41           0
              ...         ...         ...
        147944630          80           0
        147944838          80           0
        148270131          80           0
        148796126          80           0
  148808750          80           0
  ---
  seqlengths:
             10          11          12          13  ...
  104305016  107310763   91163125   84240350  ...
              5           6           7          MT           X
  121191424  119458736  112638659   16338  148823899
```

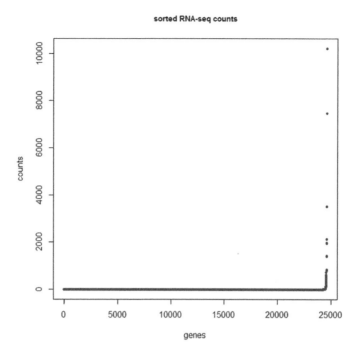

Fig. 5.13   Plot of sorted RNA-seq counts

We then use *GRanges* to rename our *reads* and ignore the strands. Finally, the *countOverlaps* function was used to create a vector with the number of alignments that correspond to each of the identifiers (*gene ids*) in *txGene*.

```
> head(counts)
```

```
ENSBTAG00000000005  ENSBTAG00000000008  ENSBTAG00000000009
                 0                   0                   0
ENSBTAG00000000010  ENSBTAG00000000011  ENSBTAG00000000012
                 2                   0                   0
```

While here there is a single sample, in practice we would create a loop to read all BAM files, rename and get the counts. Then put it all together into a matrix of counts that would be similar to the data from the arrays but with counts instead of intensities. To get an idea of the number of counts we can make a plot (Fig. 5.13)

```
> plot(sort(counts),col="blue",pch=20,
+    main="sorted RNA-seq counts",
+    xlab="genes",ylab="counts",
+    cex.main=0.9)
```

Notice how most of the genes have very low expression counts. Let's have a closer look

```
> length(counts)
```

```
[1] 24616
```

```
> length(which(counts>0))
```

```
[1] 7102
```

```
> index=which(counts>=200)
> length(index)
```

```
[1] 39
```

```
> sum(counts[index])/sum(counts)
```

```
[1] 0.554584
```

There are 24,616 known features and 7,102 are expressed at least once in this sample. There are only 39 transcripts with more than 200 counts (*length(index)*) and they account for over 55 % of the data. Finally the count data can be saved with

```
> write.table(counts,
+   "chapter5/expressionCounts.txt",
+   quote=F,sep="\t",
+   col.names=F)
```

At this point we are ready to use the RNA-seq data for differential expression analysis. No point in keeping non-expressed features, so the first step would be to remove them from the data. We only have one sample, so we cannot do an analysis but in the book chapter's folder there are two files—*RNAdat.txt* and *RNAcontrast.txt*. This is just an example with simulated data (and purposely quite bad data) for us to do a basic DE analysis; the first file is count data for 20 samples (10 controls × 10 treatments) using the same 7,102 *genes* that were expressed in the file we have worked with so far. The second file just matches samples to treatment groups and is useful to build design matrices and contrasts. The general steps for the analysis are quite similar to what we did with the arrays. First a normalization step, the total read counts are not the same for each sample and need to be adjusted. Always good to make use of plots to get a feeling for the data. Then test for differential expression and correct for multiple testing. In terms of testing, the key difference is that here we have count data—the distribution of the data is different (a negative binomial distribution is a good choice). The code below is largely based on [7] and an in-depth discussion can be found there.

A good package for DE with RNA-seq is *edgeR*; another popular one is *DESeq*. We will use the former one.

```
> library(edgeR)

> counts=read.table("chapter5/RNAdat.txt",
+   header=T,sep="\t")

> contrast=read.table("chapter5/RNAcontrast.txt",
+   header=T,sep="\t")

> DGE=DGEList(counts,group=contrast$contrast)
> DGE=calcNormFactors(DGE)
> DGE=estimateCommonDisp(DGE)
> DGE=estimateTagwiseDisp(DGE)

> # exact test
> difexpEx=exactTest(DGE,pair=c("ctrl","treat"))
> resEx=topTags(difexpEx,n=nrow(DGE))
> outEx=resEx$table
```

We loaded the *edgeR* library, then read in the data. *DGEList* is a container for RNA-seq data; we created one called *DGE* with the counts and *group* is used to assign control/treatment classes to the respective samples. Next we estimated the normalization and dispersion factors (used in the modeling to adjust for data effects). Normalization is still a topic of interest in RNA-seq analysis, with quite a few approaches to the problem (for a comparison and discussion see [23]). For further details see the *edgeR* manual or [7]. Then performed a simple contrast between the two groups (*exactTest*) and extracted results from the model using *topTags* (*n=nrow(DGE)* was used to get values for all features, but could also get only the top *n* results if desired).

The data here is quite bad. Notice in Fig. 5.14 that the coefficient of variation is quite high at low counts; the first step would be to exclude these from the data and rerun the analysis.

```
> par(mfrow=c(1,2))
> plotMeanVar(DGE,show.tagwise.vars=T,NBline=T)
> plotBCV(DGE)
```

To see the significantly differentially expressed features ($p < 0.05$) after FDR correction

```
> outEx[outEx$FDR<0.05,c(1,2,4)]
```

	logFC	logCPM	FDR
ENSBTAG00000006999	0.9884217	12.799881	4.855869e-61
ENSBTAG00000007782	0.9886040	11.962254	1.844169e-54
ENSBTAG00000007172	0.9797505	9.146706	4.465598e-18
ENSBTAG00000020125	0.9161259	8.499507	2.664356e-09
ENSBTAG00000009055	1.0132945	8.151157	1.406032e-08

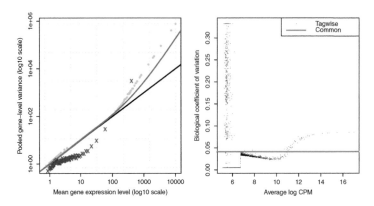

Fig. 5.14 Mean–variance relationship and biological coefficient of variation

```
ENSBTAG00000000855 1.1838970   7.739359 2.287217e-08
ENSBTAG00000017267 0.9936588   8.026673 1.618759e-07
ENSBTAG00000004910 1.0096364   7.830722 2.111096e-06
ENSBTAG00000001393 1.0311526   7.632497 1.223259e-05
ENSBTAG00000002493 1.1216396   7.385824 1.520752e-05
ENSBTAG00000012816 0.8354407   8.092320 1.520752e-05
ENSBTAG00000020630 0.9166478   7.835276 2.335346e-05
ENSBTAG00000001721 0.9187545   7.752196 5.597572e-05
ENSBTAG00000013716 1.5162717   6.476179 4.321941e-04
ENSBTAG00000001424 1.0352284   7.108584 2.774393e-03
ENSBTAG00000009949 0.9746238   7.210936 2.984333e-03
ENSBTAG00000000801 1.1651122   6.806048 3.813573e-03
ENSBTAG00000000025 0.8848511   7.318589 5.249475e-03
ENSBTAG00000006748 0.8856563   7.180246 9.298195e-03
ENSBTAG00000014448 1.5243525   6.166851 1.103187e-02
ENSBTAG00000024199 0.9480600   6.998231 1.988713e-02
ENSBTAG00000009366 1.6453899   5.946605 2.368719e-02
ENSBTAG00000012267 0.8515862   7.179946 3.391110e-02
ENSBTAG00000037972 1.4873236   5.990751 3.982897e-02
ENSBTAG00000001657 0.9060280   6.954235 4.102073e-02
ENSBTAG00000047694 0.8515470   7.068026 4.515505e-02
```

There are 26 differentially expressed features. There were 50 in the simulated data and the analysis correctly identified 22. There are 4 false-positives and 24 false-negatives but don't read much into this—the data is quite unrealistic. A similar result can be obtained by fitting a *GLM* model instead of *exactTest*. It is similar to what we did with the arrays—create a design matrix, fit a model, and test a contrast. For details on the code see [7] and the documentation of *edgeR*.

```
# glm
design=model.matrix(~ contrast, contrast)

# estimate design matrix dispersions
DGE2=estimateGLMTrendedDisp(DGE,design)
DGE2=estimateGLMTagwiseDisp(DGE2,design)

# fit model
fit=glmFit(DGE2,design)

# test contrast (coef) from design matrix
difexpGLM=glmLRT(fit,coef=2)

# same as with exactTest
res=topTags(difexpGLM,n=nrow(DGE))
outGLM=res$table
outGLM[outGLM$FDR<0.05,]
```

A last couple of notes on RNA-seq. This overview was really superficial, there is a lot more to RNA-seq than what we discussed—there is a huge number of parameters in each step that can change results. It is still relatively new and there is not much consensus about *golden standards* for the analysis. There are many steps in the analysis and each one will carry over the effects of what was done in the previous stage. Analysis are somewhat subjective and heavily reliant on external data (reference assemblies, annotation...), chances are that no two analysis will come up with the same results but hopefully they will at least be similar (the analysis is a little unique in that the actual data used changes—*counts* are always different depending on the method used). There should be more concordance with larger, clearer effects; much less so when thresholds are less well defined. But even with all this uncertainty we can still make good inroads and the best way forward is to make sure that all steps are clearly documented so they can be repeated or re-evaluated down the track. Lastly, while non-R software was used with the RNA-seq, it is still possible to use R as the primary pipeline for the analysis by calling the other programs from R and using its outputs for downstream work (see Chap. 7 for details).

5.6   Useful R Packages

The number of packages available to work with gene expression data in R is huge. I'll just list some of the more usual libraries that may be useful for your work. It's definitely worthwhile a visit to Bioconductor to find out more on what is available.

- *Biobase, affy, affxparser, marray, oligo*—back room work
- *affy, vsn, plier, affyPLM*—normalization

- *arrayQuality, simpleaffy, maDBm gplots, geneplotter, ABarray*—summaries and plotting
- *limma, GlobalAncova, genArise, pickgene*—differential expression
- *OCplus, fdrtool, multtest*—false discovery, rates multiple testing
- *ShortRead, IRanges, GenomicRanges, GenomeInfoDb, AnnotationDbi, Biostrings, BSgenome, Rsamtools*—for handling and manipulation of sequence data
- *DESeq, edgeR*—differential expression with RNA-seq

Chapter 6
Databases and Functional Information

In this chapter we discuss how to translate the statistical results from analysis into biological knowledge using online databases of genomic data and annotation packages in R. We also discuss functional profile analysis using gene ontologies and pathway analysis. Some examples for testing overrepresentation of differentially expressed genes are shown.

6.1 Introduction to Databases

We take it for granted but the internet has revolutionized the way we work. We have access to almost unlimited sources of information 24/7, all we have to do is find it (and hope that our library subscribes to the non-open access journals). The internet also enabled the existence of online repositories—vast databases of genomic data and biological knowledge. In GWAS for some applications (e.g., genomic breeding values) it might be enough to know that a certain SNP with a cryptic name is associated with a trait without any further information. For gene expression, to find out that, e.g., Affymetrix probe identifier *Bt.24953.1.S1_at* is differentially expressed is just the beginning of the story. The next step is to find out what gene this probe maps to and what the gene does, what is its biological function, what is its relevance to our problem. That's where the online databases (*DBs*) become important—we can *annotate* our probe (gene) to what is known about it: function, protein, biochemical pathway, publications that have studied this gene and so on.

There are many databases out there—there's even a journal entirely dedicated to the subject: *DATABASE* [65]. As a guide there are databases for:

Electronic supplementary material The online version of this chapter (doi: 10.1007/ 978-3-319-14475-7_6) contains supplementary material, which is available to authorized users.

© Springer International Publishing Switzerland 2015
C. Gondro, *Primer to Analysis of Genomic Data Using R*, Use R!,
DOI 10.1007/978-3-319-14475-7_6

- Literature with references (and full text) publications
- Taxonomies with classification information
- Sequence databases with DNA information
- Genomic databases with gene information
- Protein databases
- Protein families, domains, and functional sites
- Enzyme and metabolic pathways

Most DBs cross reference each other (and/or duplicate content); so it's easy to navigate from one to another. The main databases that will meet most common needs are:

- *NCBI*: http://www.ncbi.nlm.nih.gov/
- *EMBL*: http://www.embl.de/
- *DDBJ*: http://www.ddbj.nig.ac.jp

And the more specialized

- *EXPASY*: http://www.expasy.org
- *KEGG*: http://www.genome.ad.jp/kegg/kegg2.html
- *GO*: http://www.geneontology.org/

The key issue is to link our data (SNP identifiers, probe ids, sequence...) to the databases. Manufacturers will usually supply (or have available from their website) either an annotation file with identifiers (e.g., microarrays) to the various databases or genomic position information (e.g., SNP chips) or at least sequence data that will allow queries against a reference genome (e.g., primer sequence information). When the interest is genes, a good identifier is *EntrezGene*, it's a numeric identifier unique for each gene in each species—no problems with dubious meanings such as gene symbols which have aliases. *EntrezGene* works well for cross referencing, in most databases it is possible to map from *EntrezGene* to identifiers in other DBs. In practice this would mean, in the case of microarrays, mapping from the probe to its *EntrezGene* id and from there to the other database identifiers. And by the way, the *EntrezGene* identifier for probe *Bt.24953.1.S1_at* is *539076* and a search in EntrezGene (Fig. 6.1) will tell us that it is a *Bos taurus* gene known as gametogenetin binding protein 2 (GGNBP2).

6.2   Gene Annotation

Fortunately R has very good annotation options. We can interact directly with external databases via a browser, from within R itself or using R's *annotation packages*. Our focus will be on the last option. These *annotation packages* are essentially *SQLite* databases built from various public DB sources with R wrapper functions to query content. In reality the underlying process is very similar to what we did in Chap. 3: establish a connection to the DB, send an SQL query and return the data as a data.frame, vector or a list. The difference is that the wrapper functions

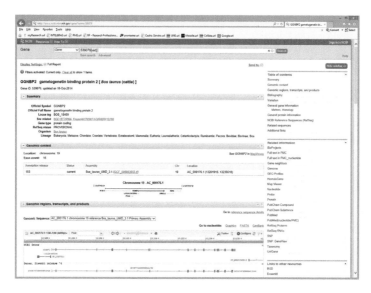

Fig. 6.1 Screenshot of NCBI's EntrezGene website

do all the hard work for us—we don't need to know anything about SQL. There are many annotation packages available with three main categories: (1) platform specific—annotation data mapping probes for a specific array such as the Affymetrix bovine chip we used in Chap. 5 to DB identifiers (e.g., *bovine.db*); (2) species specific—gene centric databases with all known genes for the species and their mapping to for instance KEGG, GO, Pubmed among others (e.g., *org.Bt.eg.db*); (3) genomic databases—genomic data for individual species, careful because they are very large downloads (e.g., *BSgenome.Btaurus.UCSC.bosTau4*). Have a look at the Bioconductor website under *Annotation Data* to see what's available. As a side note, in the same section there are also CDF packages (e.g., *bovinecdf*) for the Affymetrix arrays and probe sequence data for many arrays (e.g., *bovineprobe*). I made a point of giving an example of each one because the naming conventions are pretty well structured; it's quite easy to find what you are looking for.

Before delving into the *annotation packages* let's have a quick look at the first two options to access external DBs. The hard way is to know how to query the DB via a browser. For example using our *539076—GGNBP2* we could type

```
> Entrezquery=
+       "http://www.ncbi.nlm.nih.gov/gene?term="
> browseURL(paste(Entrezquery,"539076",sep=""))
```

Which would open the default browser with exactly the same results shown in Fig. 6.1. The problem is that you need to know how to query each database. Hint: normally you can find this out from the database itself looking at the url address, just type in a query and see what the address looks like. In the package *annotate* you will also find some information on how to query the DBs and there are some ready made functions as well. For example

```
> library(annotate)
> browseURL(pmidQuery(18648396))
```

will directly open a browser window for the pubmed identifier *18648396*). All
pmidQuery does is what we did above, paste a url to an identifier (see also *Uni-
GeneQuery*). Another hint: type in the name of the function (e.g., *UniGeneQuery*)
and you can see how the query is setup in case you want to do it manually.

This is handy sometimes but not overly efficient. A better way is with BioMart
which establishes a direct connection to databases which use the system (for details
see http://www.biomart.org).

You'll need the library *biomaRt* and then you can create a mart connection from
one of the available marts.

```
> library(biomaRt)
> listMarts()[,1]

 [1] ensembl
 [2] snp
 [3] functional_genomics
 [4] vega
 [5] fungi_mart_22
 [6] fungi_variations_22
 [7] metazoa_mart_22
 [8] metazoa_variations_22
 [9] plants_mart_22
[10] plants_variations_22
[11] protists_mart_22
[12] protists_variations_22
[13] msd
[14] htgt
[15] REACTOME
[16] WS220
[17] biomart
[18] pride
[19] prod-intermart_1
[20] unimart
[21] biomartDB
[22] biblioDB
[23] Eurexpress Biomart
[24] phytozome_mart
[25] metazome_mart
[26] HapMap_rel27
[27] cildb_all_v2
[28] cildb_inp_v2
[29] experiments
```

```
[30] oncomodules
[31] europhenomeannotations
[32] ikmc
[33] EMAGE gene expression
[34] EMAP anatomy ontology
[35] EMAGE browse repository
[36] GermOnline
[37] Sigenae_Oligo_Annotation_Ensembl_61
[38] Sigenae Oligo Annotation (Ensembl 59)
[39] Sigenae Oligo Annotation (Ensembl 56)
[40] Breast_mart_69
[41] K562_Gm12878
[42] Hsmm_Hmec
[43] Pancreas63
[44] Public_OBIOMARTPUB
[45] Public_VITIS
[46] Public_VITIS_12x
[47] Prod_WHEAT
[48] Public_TAIRV10
[49] Public_MAIZE
[50] Prod_POPLAR
[51] Prod_POPLAR_V2
[52] Prod_BOTRYTISEDIT
[53] Prod_
[54] Prod_SCLEROEDIT
[55] Prod_LMACULANSEDIT
[56] vb_mart_24
[57] vb_snp_mart_24
[58] expression
[59] ENSEMBL_MART_PLANT
[60] ENSEMBL_MART_PLANT_SNP
```

listMarts() shows the available DBs. Let's connect to *ensembl* and use the bovine dataset.

```
> amart=useMart("ensembl","btaurus_gene_ensembl")
```

To find out what datasets are available for a given database use *listDatasets (amart)*. Now let's query the DB for our probe *Bt.24953.1.S1_at*

```
> getBM(attributes="entrezgene",
+      filters="affy_bovine",
+      values="Bt.24953.1.S1_at",
+      mart=amart)
  entrezgene
1    539076
```

```
> getBM(attributes="family_description",
+       filters="affy_bovine",
+       values="Bt.24953.1.S1_at",
+       mart=amart)

     family_description
1 GAMETOGENETIN BINDING 2
```

The function *getBM* is the main query function. You'll need to define as arguments
what you want to query (*attributes*)—in our example we searched for the entrez
identifier which, as expected, returned the same *539076* we had before; and we also
searched for *family_description* which returned the same results we saw in Fig. 6.1.
We also have to specify a filter (*filters*)—what kind of information we are supplying
(here affymetrix ids), then the actual query (*values*)—here our probe id (note that
more than one can be queried each time, the same for attributes, we did not need to
make two independent queries) and finally we supply our mart connection.

To find out what attributes and filters are available for the DB use *listAt-
tributes(amart)* and *listFilters(amart)*. A word of caution, you might run into trouble
to get through the firewall from your institution since the ports used by BioMart
may be blocked. Try to change the port using a MySQL connection (see details in
the *useMart* help files).

Now back to the *annotation packages*. Their advantage is that they are on
the local machine, no struggling with firewalls (nor sys admin wardens) and no
connection latencies. But they can get a bit outdated between releases and of course
there has to be a package for your platform or at least your organism (if not have
a look at *AnnotationDbi, annotationTools, Resourcerer...*). Alternatively, setup a
local BioMart database that you can easily update yourself.

Let's continue with the bovine Affymetrix array. The library is *bovine.db*. Start
by loading it and let's see what's in it.

```
> library(bovine.db)
> bovine()

Quality control information for bovine:

This package has the following mappings:

bovineACCNUM has 24117 mapped keys (of 24128 keys)
bovineALIAS2PROBE has 14028 mapped keys (of 41458 keys)
bovineCHR has 16565 mapped keys (of 24128 keys)
bovineCHRLENGTHS has 11692 mapped keys (of 11692 keys)
bovineCHRLOC has 13455 mapped keys (of 24128 keys)
bovineCHRLOCEND has 13455 mapped keys (of 24128 keys)
bovineENSEMBL has 15597 mapped keys (of 24128 keys)
bovineENSEMBL2PROBE has 10755 mapped keys (of 18746 keys)
bovineENTREZID has 16579 mapped keys (of 24128 keys)
bovineENZYME has 2371 mapped keys (of 24128 keys)
```

```
bovineENZYME2PROBE has 821 mapped keys (of 953 keys)
bovineGENENAME has 16579 mapped keys (of 24128 keys)
bovineGO has 6411 mapped keys (of 24128 keys)
bovineGO2ALLPROBES has 11528 mapped keys (of 12179 keys)
bovineGO2PROBE has 7228 mapped keys (of 7813 keys)
bovinePATH has 5673 mapped keys (of 24128 keys)
bovinePATH2PROBE has 224 mapped keys (of 225 keys)
bovinePFAM is deprecated because up to date
                IPI IDs are no longer available.
Please use select() if you need access to
                PFAM or PROSITE accessions.

bovinePFAM has 12540 mapped keys (of 24128 keys)
bovinePMID has 15667 mapped keys (of 24128 keys)
bovinePMID2PROBE has 7020 mapped keys (of 8218 keys)
bovinePROSITE is deprecated because up to date
                IPI IDs are no longer available.
Please use select() if you need access to
                PFAM or PROSITE accessions.

bovinePROSITE has 12540 mapped keys (of 24128 keys)
bovineREFSEQ has 16527 mapped keys (of 24128 keys)
bovineSYMBOL has 16579 mapped keys (of 24128 keys)
bovineUNIGENE has 14297 mapped keys (of 24128 keys)
bovineUNIPROT has 13028 mapped keys (of 24128 keys)

Additional Information about this package:

DB schema: BOVINECHIP_DB
DB schema version: 2.1
Organism: Bos taurus
Date for NCBI data: 2014-Mar13
Date for GO data: 20140308
Date for KEGG data: 2011-Mar15
Date for Golden Path data: 2012-Jun11
Date for Ensembl data: 2014-Feb26
```

Quite a lot of information. At the bottom we can see where the data came from and when it was downloaded. The *mappings* section shows the functions available to query the DB. Since it is a chip specific DB it is easy to use the Affymetrix identifiers. Another side note here, if there is no specific package for your platform use the organism DB, the array manufacturer will have some sort of annotation which can be used to cross reference to the organism's genomic DB. If you really have nothing except probe sequences try to Blast them against NCBI (but don't run a large number of automated Blast searches on NCBI, instead download the whole data from NCBI's website and run local Blast searches). Again using the Affymetrix *Bt.24953.1.S1_at* identifier to exemplify

```
> bovineENTREZID$"Bt.24953.1.S1_at"
```

```
[1] "539076"
```

```
> bovineGENENAME$"Bt.24953.1.S1_at"
```

```
[1] "gametogenetin binding protein 2"
```

```
> bovineSYMBOL$"Bt.24953.1.S1_at"
```

```
[1] "GGNBP2"
```

Yes, still the same. We can also use *mget* to query the DB.

```
> mget("Bt.24953.1.S1_at",bovineENTREZID)
```

```
$Bt.24953.1.S1_at
[1] "539076"
```

mget will accept multiple probes per query. We can also extract all data at once

```
> geneids=unlist(as.list(bovineENTREZID))
> length(geneids)
```

```
[1] 24128
```

```
> geneids[10000:10005]
```

```
   Bt.2173.1.S1_at    Bt.21730.1.S1_at Bt.21730.2.S1_a_at
         "532724"           "505288"           "505288"
  Bt.21731.1.S1_at Bt.21731.2.S1_a_at   Bt.21732.1.S1_at
         "505406"           "505406"           "508245"
```

As we would expect there's one entry for each probe. This can be useful to see for example if the probes only match a single gene or if there are more than one probe per gene.

```
> length(unique(geneids))
```

```
[1] 11283
```

Less than half! I wonder what's the maximum number of probes that a single gene maps to.

```
> summary(factor(geneids))[1]
```

```
785621
    33
```

Thirty three. That's a lot. Let's find out some more about the gene.

```
> Affyid=
+       names(geneids)[which(geneids=="785621")[1]]
> print(Affyid)

[1] "Bt.13003.1.A1_at"

> mget(Affyid,bovineGENENAME)

$Bt.13003.1.A1_at
[1] "T cell receptor, alpha"

> mget(Affyid,bovineSYMBOL)

$Bt.13003.1.A1_at
[1] "TCRA"
```

Since the *annotation packages* are just SQL relational databases we can also query them directly as we did in Chap. 3. For more details see [47]. These databases are not only useful in R but can also be used for other applications—simply navigate to the folder with the database (a folder inside the package under *library* in R) and connect to it via *SQLite*. Figure 6.2 shows SQLiteStudio (excellent viewer for SQLite databases) connected to the bovine database (sometimes it is also easier to simply look at the database's contents to understand it's structure).

Before moving onto the next topic—enrichment analysis, let's quickly see how to map the results from a GWAS study and the RNA-seq data from the previous chapter to functional information.

For argument's sake, let's say that a recent study found that SNP *BTB-01143580* is highly associated with body weight in cattle. To identify candidate genes in this region, we first need to know the location of the SNP and then search for genes in this region. We start by reading in the map information

```
> map=read.table("chapter6/map.txt",header=T,sep="\t")
> target=map[which(map$snp=="BTB-01143580"),]$position
> map[which(map$snp=="BTB-01143580"),]

              snp chrom position
23538 BTB-01143580    14 24383627
```

The variable target holds the position on chromosome 14 (now we know where the SNP is). The next step is to build a data.frame with the position of all known genes on chromosome 14 using the genomic database (this one uses gene ids as the primary key and is not platform dependent).

```
> library(org.Bt.eg.db)

> genepos=org.Bt.egCHRLOC
> mapped=mappedkeys(genepos)
> genepos = as.list(genepos[mapped])
```

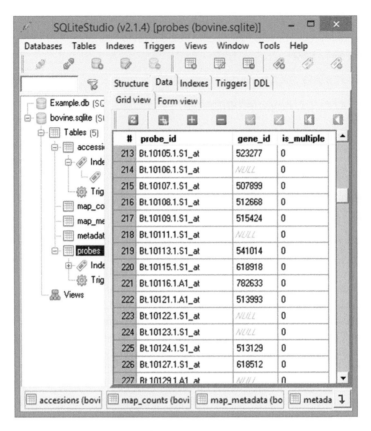

Fig. 6.2 Affymetrix bovine database in SQLiteStudio

```
> pos=matrix(NA,length(genepos),3)
> for (i in 1:length(genepos))
+ {
+   pos[i,1]=names(genepos[i])
+   pos[i,2]=names(genepos[[i]][1])
+   pos[i,3]=genepos[[i]][1]
+ }
> pos=pos[which(pos[,2]=="14"),]
> colnames(pos)=c("gene","chr","position")
> pos=as.data.frame(pos)
> pos$position=as.numeric(as.character(pos$position))
> pos$position=abs(pos$position)
```

```
> head(pos)
       gene chr position
1 100124522  14 67591848
2 100125936  14 48259383
3 100126183  14 19653745
4 100138046  14   878768
5 100140032  14 34007290
6 100196897  14 75741668
```

First we extracted the chromosome locations from the database (*CHRLOC*); then iterated through the list to build a data.frame with *gene ids*, *chromosome*, and *position*. Since the only interest here is chromosome 14, the rest was discarded. A bit of fixing up of factors into numeric (the positions) and making them all positive (the minus sign is for the opposite strand).

Now it is simply a matter of finding genes around this SNP. Here we are just getting the closest one but normally we would look into a window (e.g., up and downstream 0.5 Mbp). There are many different ways of doing this, a simple way is to just get the absolute differences between *target* and *position* and then match it to the gene, e.g.

```
> dists=pos$position-target
> mindist=min(abs(dists))

> index=which(dists==mindist | dists== -mindist)
> pos[index,]

       gene chr position
167 530316  14 24359901
```

And the gene is 530316. To get some information about it

```
> mget(as.character(pos[index,1]),org.Bt.egGENENAME)
$`530316`
[1] "UBX domain protein 2B"

> browseURL(paste(Entrezquery,pos[index,1],sep=""))
```

This shows it is quite simple to match SNP with gene information. A package that can help with this is *NCBI2R*. Lastly, let's match the RNA-seq data to gene ids. Read in the data from the previous chapter; the *rownames* are the only information we have to work with.

```
rna=read.table("chapter6/RNAdat.txt",header=T,sep="\t")
rna=rownames(rna)
head(rna)

[1] "ENSBTAG00000000010" "ENSBTAG00000000013"
[3] "ENSBTAG00000000014" "ENSBTAG00000000019"
[5] "ENSBTAG00000000022" "ENSBTAG00000000023"
```

rna holds the identifiers we want to match against *org.Bt.eg*

```
> acc=org.Bt.egENSEMBL2EG
> mapped=mappedkeys(acc)
> hold=intersect(rna,mapped)
> index=match(hold,rna)
> rna=rna[index]
> rnaToGene=mget(rna,org.Bt.egENSEMBL2EG)

> mget(as.character(rnaToGene[1]),org.Bt.egGENENAME)
$`511289`
[1] "ubiquitin-like 7 (bone marrow stromal cell-derived)"
```

It is very similar to what we did before. Basically extract all mapped *ensembl* keys; match them to the RNA-seq ids (some are not annotated, so were excluded) and then get a list of transcript ids as the names of the list items and gene id (or ids) as elements. To get a gene description, *mget* was used (same as with the SNP).

6.3   Gene Ontology

Gene ontologies (GO) provide a rigid controlled vocabulary to describe genes and gene products in various species. Ontologies are structured as *directed acyclic graphs*, in practice this means a tree-like structure with more general terms on top and more specific terms below. The most general terms in GO are *molecular function*—what a gene does at the biochemical level; *biological process*—which process a gene is involved in; and *cellular component*—where a gene is found within cells or complexes [18].

The main uses of GO are twofold: help reduce the dimensionality of results and look for overrepresentation of specific terms and consequently functional overrepresentation. In the first case, we might come up with a list of 500 DE genes, it's very hard to go over each gene individually, see what each one does and try to make sense out of it all. GO terms can help to group these genes into clusters with similar properties making the analysis much more tractable. In the second case we extend this concept of grouping a bit further and formally test if the genes in our result are more frequent than we would expect by chance, this is an indication that these overrepresented terms are involved in our trait of interest. A nice side effect of GO analysis is that it can help reduce the noise (false positives) in expression experiments, instead of focusing on individual genes we focus on processes. An example of the use of GO in expression analysis is given in [87].

To illustrate let's take a list of probes from the file *DEprobes.txt* which is in the folder for this chapter. Here we are again using Affymetrix array probes but if working with RNA-seq data it is simply a matter of using the organism's database instead (as above for mapping RNA-seq data). What does the annotation package tell us about them?

Fig. 6.3 Screenshot of Gene Ontology website

```
> deprobes=read.table("chapter6/DEprobes.txt",
+       header=T,sep="\t")
> goannot=toTable(bovineGO[as.vector(deprobes$AFFY)])
> head(goannot)
```

```
          probe_id       go_id Evidence Ontology
1 Bt.11558.1.A1_at GO:0001666      IEA       BP
2 Bt.11558.1.A1_at GO:0001934      IEA       BP
3 Bt.11558.1.A1_at GO:0007005      IEA       BP
4 Bt.11558.1.A1_at GO:0007015      IEA       BP
5 Bt.11558.1.A1_at GO:0007021      IEA       BP
6 Bt.11558.1.A1_at GO:0007601      IEA       BP
```

toTable is a handy function to structure the results from the DB. Take the first term from our results *GO:0001666* and let's see what it is (we already know it's a BP—*biological process*, the other two shorthands are *MF* and *CC*). There's another column in the table called *Evidence*. These are evidence codes which give an indication of how *confident* we are about the term. *IEA* means *inferred from electronic annotation*. Check the guide to evidence codes on the GO site for other codes. A search on the GO website will tell us that the term is associated with response to hypoxia (Fig. 6.3). The same information can be obtained in R using the annotation package *GO.db* which is specific for gene ontologies.

```
> library(GO.db)
> GO()
```

```
Quality control information for GO:


This package has the following mappings:

GOBPANCESTOR has 25683 mapped keys (of 25683 keys)
GOBPCHILDREN has 14773 mapped keys (of 25683 keys)
```

```
GOBPOFFSPRING has 14773 mapped keys (of 25683 keys)
GOBPPARENTS has 25683 mapped keys (of 25683 keys)
GOCCANCESTOR has 3441 mapped keys (of 3441 keys)
GOCCCHILDREN has 1152 mapped keys (of 3441 keys)
GOCCOFFSPRING has 1152 mapped keys (of 3441 keys)
GOCCPARENTS has 3441 mapped keys (of 3441 keys)
GOMFANCESTOR has 9679 mapped keys (of 9679 keys)
GOMFCHILDREN has 1963 mapped keys (of 9679 keys)
GOMFOFFSPRING has 1963 mapped keys (of 9679 keys)
GOMFPARENTS has 9679 mapped keys (of 9679 keys)
GOOBSOLETE has 1860 mapped keys (of 1860 keys)
GOTERM has 38804 mapped keys (of 38804 keys)


Additional Information about this package:

DB schema: GO_DB
DB schema version: 2.1
Date for GO data: 20140308

> mget(goannot[1,2],GOTERM)

$`GO:0001666`
GOID: GO:0001666
Term: response to hypoxia
Ontology: BP
Definition: Any process that results in a change in
  state or activity of a cell or an organism
  (in terms of movement, secretion, enzyme production,
  gene expression, etc.) as a result of a stimulus
  indicating lowered oxygen tension. Hypoxia, defined
  as a decline in O2 levels below normoxic levels
  of 20.8--20.95%, results in metabolic adaptation at
  both the cellular and organismal level.
Synonym: response to hypoxic stress
Synonym: response to lowered oxygen tension
```

GOTERM gives a general description of the term (compare to the top of Fig. 6.3). Recall that GO is a tree like structure. The immediate term above the term and all higher order terms can be obtained with

```
> mget(goannot[1,2],GOBPPARENTS)

$`GO:0001666`
        is_a            is_a
"GO:0006950" "GO:0036293"
```

```
> mget(goannot[1,2],GOBPANCESTOR)
```

```
$`GO:0001666`
[1] "GO:0006950" "GO:0008150" "GO:0009628" "GO:0036293"
[5] "GO:0050896" "GO:0070482" "all"
```

Note the use of *BP* because we are dealing with a biological process. To work down the tree use *CHILDREN* and *OFFSPRING*.

Next we can test for overrepresentation using the package *topGO*. When testing for overrepresentation the key point is to make sure that a valid *universe* is used—this is the set of all possible probes or genes that the sampling came from. For example, don't use all known genes for an organism if the array you used only had a proportion of these; with RNA-seq preferably use the list of genes that had some level of expression in the samples. Also if some pre-filtering was conducted, use only the features that were actually used in testing for differential expression. Here for simplicity we will (wrongly) use all probes on the Affymetrix bovine array. First extract the probe names from the DB then create a vector of class factor with *0/1* for all probes, *0* for not DE and *1* for DE, then name the vector with the probe names so we know what is what. The next step is to create an object of class *GOdata* for *molecular function, MF* (see syntax details and information on *GOdata* objects in the *topGO* help files).

```
> library(topGO)
> probes=
+     names(unlist(as.list(bovineENTREZID)))
> genelist =
+     factor(as.integer(probes %in% deprobes$AFFY))
> names(genelist) = probes
> GOdata = new("topGOdata", ontology = "MF",
+     allGenes = genelist, annot = annFUN.db,
+     affyLib = "bovine.db")

Building most specific GOs .....
( 1885 GO terms found. )

Build GO DAG topology .........
( 2470 GO terms and 3080 relations. )

Annotating nodes ..............
( 5258 genes annotated to the GO terms. )
```

Once we have a *GOdata* object it is very easy to test terms for overrepresentation. Select a test and then test it on the *GOdata* object.

```
> test = new("classicCount",
+     testStatistic = GOFisherTest,
+     name = "Fisher test")
> resultFis = getSigGroups(GOdata, test)
```

```
                        -- Classic Algorithm --

                the algorithm is scoring 173 nontrivial
                nodes test statistic:  Fisher test
```

score(resultFis) will return the *p*-values for each GO term that was tested. An
easier way to see results is with

```
> GenTable(GOdata,Fisher=resultFis,topNodes=10)
```

```
        GO.ID
1   GO:0005200
2   GO:0005344
3   GO:0019825
4   GO:0043425
5   GO:0001102
6   GO:0070888
7   GO:0016757
8   GO:0016758
9   GO:0008375
10  GO:0033613
                                                Term
1           structural constituent of cytoskeleton
2                         oxygen transporter activity
3                                      oxygen binding
4                 bHLH transcription factor binding
5   RNA polymerase II activating transcripti...
6                                      E-box binding
7   transferase activity, transferring glyco...
8   transferase activity, transferring hexos...
9         acetylglucosaminyltransferase activity
10      activating transcription factor binding
    Annotated Significant Expected  Fisher
1            13            3        0.08 5.1e-05
2            14            3        0.08 6.5e-05
3            18            3        0.11 0.00014
4             7            2        0.04 0.00069
5             8            2        0.05 0.00092
6             8            2        0.05 0.00092
7            81            4        0.48 0.00120
8            60            3        0.35 0.00506
9            19            2        0.11 0.00540
10           19            2        0.11 0.00540
```

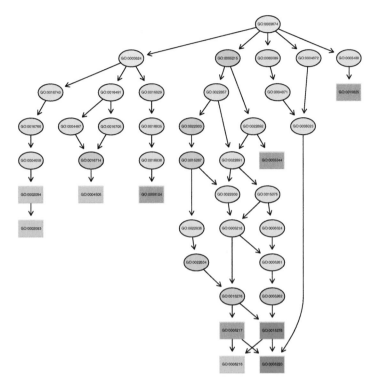

Fig. 6.4 Tree plot of top ten overrepresented *molecular function* terms

This returns the top most significant GO terms for *molecular function*. We can also plot (Fig. 6.4) and save these terms using

```
> showSigOfNodes(GOdata, score(resultFis),
+       firstSigNodes = 10)

$dag
A graphNEL graph with directed edges
Number of Nodes = 28
Number of Edges = 31

$complete.dag
[1] "A graph with 28 nodes."
```

Figure 6.4 shows the relationships between the nodes; red squares are the more significant terms. We can do exactly the same with RNA-seq and even significant SNP identified by GWAS for which a list of neighboring genes was made (as we did above). For more details on GO start with [47], [40] and [6].

6.4   Pathway Analysis, Physical Mapping, and Protein Domains

Other common *enrichment* analyses are based on: (1) KEGG, the Kyoto Ency-
clopedia of Genes and Genomes which is a database that contains information
about metabolic pathways in many organisms; (2) on chromosome locations based
on physical mapping information; and (3) on protein domain information (e.g.,
PROSITE). All of these have similar objectives to what we did in the previous
section—identify overrepresented features in a pathway, in a chromosome region, in
a domain. We will not go into details herein, some worked out examples are given
in [47] and [40]. Useful R packages are mentioned in the next section. To briefly
illustrate we could use functions from the library *Category* to test our list of probes
for overrepresentation. Initially we create a list of unique gene identifiers for our DE
probes (*keggDE*) and for our *universe* (*keggU*). Next we build the parameters for a
hypergeometric test and then test using the function *hyperGTest* (see the help files
for syntax details). We can see the pathways with a *p*-value < 0.01 using *summary*.
Figure 6.5 shows a pathway (not from this example) with non-genic elements in
yellow, non-DE genes in gray, and the DE genes colored in red or green according
to the fold change (positive and negative, respectively).

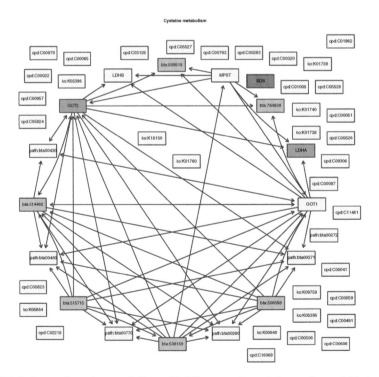

Fig. 6.5 Pathway of cysteine metabolism with DE genes color coded according to fold change

```
> library(Category)
> keggDE=unique(toTable(bovineENTREZID
+      [as.vector(deprobes$AFFY)])$gene_id)
> keggU=unique(toTable(bovineENTREZID
+      [as.vector(probes)])$gene_id)
> params=new("KEGGHyperGParams",geneIds=keggDE,
+      universeGeneIds=keggU,annotation="bovine",
+      pvalueCutoff=0.01,testDirection="over")
> keggtest=hyperGTest(params)
> summary(keggtest,pvalue=0.01)
```

	KEGGID	Pvalue	OddsRatio	ExpCount	Count	Size
1	05310	0.0006823845	21.140977	0.1795209	3	22
2	04612	0.0007475868	11.481481	0.4243222	4	52
3	05332	0.0015576683	15.420330	0.2366412	3	29
4	04145	0.0019502805	6.517094	0.9220848	5	113
5	05330	0.0020796666	13.814039	0.2611213	3	32
6	05320	0.0022750406	13.350000	0.2692814	3	33
7	04940	0.0031692115	11.766807	0.3019216	3	37
8	05323	0.0032409435	7.498732	0.6283232	4	77
9	04672	0.0034219349	11.427551	0.3100816	3	38
10	05150	0.0048698682	9.985714	0.3508818	3	43
11	05145	0.0074054439	5.854241	0.7915241	4	97
12	05416	0.0078636943	8.303571	0.4161621	3	51

	Term
1	Asthma
2	Antigen processing and presentation
3	Graft-versus-host disease
4	Phagosome
5	Allograft rejection
6	Autoimmune thyroid disease
7	Type I diabetes mellitus
8	Rheumatoid arthritis
9	Intestinal immune network for IgA production
10	Staphylococcus aureus infection
11	Toxoplasmosis
12	Viral myocarditis

The objective of this chapter was to provide a basic overview of how to explore functional knowledge in the context of genomic analysis. This is just the tip of the iceberg; as more knowledge is created and stored in computationally retrievable form we can expect that embedding prior knowledge into the analytical steps will become more prevalent. For *enrichment* analysis, Bioconductor is again the primary repository to investigate. Here we focused on functional annotation databases, but there are many other sources of information for sequence data as well that can be used in conjunction with the RNA-seq data from the previous chapter.

6.5   Useful R Packages

- Annotation packages—*GO.db, KEGG.db, reactome.db* and platform or species specific DBs (note that KEGG.db is no longer being updated—better avoid)
- *AnnotationDbi, annaffy, annotationTools, Resourcerer*—annotation tools
- *Category, GlobalAncova*—general analysis
- *GOstats, goTools, topGO, goProfiles*—GO
- *KEGGgraph, PGSEA, GSEABase, KEGGREST*–KEGG
- *macat*—chromosome location
- *domainsignatures*—protein domains
- *Rgraphviz, geneplotter, GenomeGraphs*—plotting

Chapter 7
Extending R

In this chapter we will overview some additional options to work with R: how to speed up computations and better ways to handle data. Simple parallelization (and pseudo-parallelization) is discussed along with some packages for R. Sometimes additional programs are needed for an analysis, we will see how to interface with them and also how to write programs in other languages for use in R. Many applications need a graphical interface, we will illustrate how to build graphic shells and use R as the *engine* behind the scenes. Results from an analysis are of limited value unless they are reproducible and reported in a *human digestible* format—we will see some of R's reporting functionalities.

7.1 Large Data–Large Problems

Moore's law has been holding well for computer hardware: chip performance doubles roughly every couple of years. The drops in the cost of sequencing over time closely reflected Moore's law until around 2008. With the advent of next-gen sequencing platforms, costs started to fall at a much faster pace than Moore's law. This is rapidly leading to data outstripping computational resources. At the end of the day, large projects require high-end computational resources and there is no real workaround. There are however ways to improve R's performance when working with large datasets and make better use of computational resources. In this chapter we will discuss some tips on how to improve performance in R and to how exploit the parallel architecture available in current computers. We will then see how to use other programs directly from within R and, in broad terms, mention how to write code in $C++$ that can interface with R. The coin is then flipped and we discuss a

Electronic supplementary material The online version of this chapter (doi: 10.1007/ 978-3-319-14475-7_7) contains supplementary material, which is available to authorized users.

© Springer International Publishing Switzerland 2015
C. Gondro, *Primer to Analysis of Genomic Data Using R*, Use R!,
DOI 10.1007/978-3-319-14475-7_7

simple example of how to *hook* R into a graphical user interface in C# so that R can be used as the *engine* running the computation. The chapter concludes with an overview of how to automate reporting in R. Parts of this chapter are based on [44] with permission from the publisher.

7.2   Improving Read and Write Operations in R

We already discussed *vectorization* and how important it is for speeding up R in Chap. 4—this should always be the first focus of attention to optimize your work. Aside from this, the first bottleneck commonly encountered when working with genomic data is getting it into R. Datasets tend to be rather large and it can take a significant amount of time just to load it in. Ideally data would be in binary format with a specific serialization/deserialization available for storing and retrieving the data (as for example the Affymetrix CEL files used in Chap. 5 or the *RDS* files in Chap. 4). More commonly however, data is stored as plain text files in somewhat variable formats, requiring flexibility in how data is read into R. We will focus on this latter scenario.

Up to now we have mostly relied on *read.table* to import flat files into R. It is the slowest but most flexible method. *read.table* performs many additional tasks (e.g., convert to factors, numeric, set missing values, check names, etc.). All this adds considerable overhead to the process. There are some options to speed up *read.table* if you tell R what data to expect. For example, predefine the column classes (what kind of data to expect in each column); define the number of rows in the table beforehand to avoid having to reallocate memory; and if there are no comments in the data, set the comment argument to empty. The first option can improve the read operation quite significantly; the others are more marginal. The *read.table* function works much better when there are more rows than columns. To illustrate we will read in a small file of genotypes (*geno.txt*) with 1,000 samples (rows) × 50,000 SNP (columns) and then its transpose (*tgeno.txt*) with more rows (50,000) than columns (1,000).

```
> system.time(
+ test <- read.table("chapter7/geno.txt", sep="\t")
+ )

   user   system elapsed
  67.53    0.17   67.71

> system.time(
+ test <- read.table("chapter7/geno.txt",sep="\t",
+    nrows=1000,
+    comment.char = "",
+    colClasses=c("character",rep("numeric",50000)))
+ )
```

```
    user   system elapsed
   63.19     0.09   63.29

> system.time(
+ test <- read.table("chapter7/tgeno.txt", sep="\t")
+ )

    user   system elapsed
   15.60     0.06   15.66
```

The function *system.time* is useful to benchmark runtimes. Note the use of $< -$ instead of $=$ inside the function. This is one of the rare occasions where the $=$ sign does not work. By adding the additional parameters to *read.table* we gained 4.5 s. Note that with *colClasses* the column names have to be included as well—that is why there are 50,001 column definitions instead of 50,000 and the first one is a *character*. There was some speed improvement but a really large gain is when the data is stored as *SNP* × *sample* instead of *sample* × *SNP*, the read time comes down to only 15.6 s which is over four times faster than the first two examples. As a rule try to get the data in a format that has more rows than columns, but unfortunately this is the opposite of what most other programs use. If the file has only genotypes (no row and column names), the scan function can be used instead.

```
> system.time(
+ test <- matrix(scan(file="chapter7/genoNumeric.txt",
+         what=integer(),sep="\t"),50000,1000,byrow=T)
+ )

Read 50000000 items
    user   system elapsed
    9.61     0.11    9.72
```

The *scan* function is much faster than *read.table* but simply returns a vector of items that then need to be *shaped* into the desired format. The argument *what* defines the data type and *sep* is the separator. The *matrix* function was then used to *bend* the vector into a matrix, here with 50,000 rows and 1,000 columns. Be careful with the *byrow* argument, it defines how the matrix is filled in [either by row or by column (for the latter use *byrow=F*)]. With *scan* it does not matter if the data has more columns than rows—read times are the same. If there are row and column names in the data, you will have to use *what = character()* and then do some manipulation to shape the matrix.

```
> T1=Sys.time() # keep track of how long it takes
> test = scan(file="chapter7/tgeno.txt",
+         what=character(),sep="\t")
> cnames = test[1:1000]
> test = test[-c(1:1000)]
> test = matrix(test,50000,1001,byrow=T)
```

```
> rnames = test[,1]
> test = test[,-1]
> test = matrix(as.numeric(test),50000,1000,byrow=F)
> rownames(test) = rnames
> colnames(test) = cnames
> rm(rnames,cnames)
> print(Sys.time()-T1) # show runtime
```

```
Time difference of 23.33781 secs
```

This is rather convoluted. The data is read in as a vector with *scan*, then the first 1,000 items are stored to use as column names (in *cnames*), next we remove them from the vector; bend the vector into a matrix with an additional column for the SNP names; store these in *rnames* and delete the first column from the matrix. Now we have the matrix of genotypes but it is in *character* format, so we have to convert to numeric using *as.numeric*; but this again returns a vector which has to be bent back into a matrix. Finally add the row and column names to the matrix of genotypes (*test*). It is a lot of work and takes 23 s (note the different way to measure time using *Sys.time*). This is slower than simply using *read.table* (15.6 s) and much slower than using *scan* with only numeric genotypes (9.6 s). However if the data is in *sample × SNP* format as in the first example, it is still faster (it takes over a minute to read in the data with *read.table*).

An even faster way to get data into R is with *readLines*

```
> system.time(
+ test <- readLines(con="chapter7/tgeno.txt")
+ )
```

```
   user  system elapsed
   3.29    0.08    3.37
```

and this brings it down to 3.29 s. But the downside is that *readLines* returns a plain vector of unparsed character lines which will require even more manipulation. Finally, the fastest way to get data into R is with *readChar*. This loads the whole file as a raw unparsed string. This is extremely fast and takes only 0.28 s to run (on par with, e.g., *C++*), but again without any structure to the data and you need to know beforehand the size of the data file (can use the argument *size* from the *file.info* function for this). The additional parsing steps required to convert the data to matrix format can make the whole process even slower than simply using *read.table*, but it is useful sometimes.

```
> system.time(
+ test <- readChar(con="chapter7/tgeno.txt",
+       nchars=file.info("chapter7/tgeno.txt")$size,
+       useBytes=T)
+ )
```

```
user   system elapsed
0.28     0.03     0.31
```

If the plan is to load the genotypes (or any other large dataset) more than once into R, it is worthwhile to load them once using one of the previously described approaches and then save it in a binary format. This makes the file much smaller and is quick to read in. A simple solution is to use the *save* function (but first get the genotypes in matrix format again)

```
> test <- matrix(scan(file="chapter7/genoNumeric.txt",
+       what=integer(),sep="\t"),50000,1000,byrow=T)
> save (test,file="chapter7/genotypes.bin")
```

The binary file is much smaller than the original file (98 MB versus 18 MB) and it takes only 0.72 s to load, in the correct matrix format, with

```
> system.time(
+ load ("chapter7/genotypes.bin")
+ )

user   system elapsed
0.72     0.03     0.75
```

A problem with this is that R will load the data into the workspace with the same original variable name used when the data was saved (*test*, in this case). If the variable already exists it will be overwritten. A couple of notes on *save*: it is not restricted to a single variable—use a list of variable names to store multiple objects in a single file; while binary formats are convenient, there is always some risk of changes to the serialization and the data may become unreadable in newer versions of R—flat files are still a safer way for long-term data storage. A more *elegant* way of storing (single) variables for use in R is with *saveRDS*; and to read in the data use *readRDS*. This is probably the most convenient way to work with large datasets in R (we have already used RDS files in Chap. 4).

```
> saveRDS(test,"chapter7/genotypes.rds")

> system.time(
+ newTest <- readRDS("chapter7/genotypes.rds")
+ )

user   system elapsed
0.70     0.03     0.73
```

This takes only 0.7 s to read the data in. With *RDS* we can assign the binary files to new variable names (same way as with *read.table*) and there is no risk of overwriting objects.

Apart from R's built in possibilities, a few packages that are worth exploring for GWAS data manipulation are *sqldf*, *ncdf* and *GWASTools*.

In summary, we started with a load time of more than 1 min and ended with less than 1 s. We have achieved this by using different functions and preparing the dataset in a format that is more digestible to R; the compromise is that the faster the data gets imported, the less useful the format becomes. The trade-off is speed to read in the data versus the additional overhead in later stages to parse it. If it is doable, get the data as clean as possible (no mixing of genotypes with phenotypes, no row and column names—store these separately), it will be much faster and easier to import the data into R. For repeated use of the data, binary formats can save a lot of time (and disk space).

7.3 Byte-Code Compiler

A byte-code compiler developed by Luke Tierney has been available for R since version 2.14 and most packages are already making use of it. The package itself (*compiler*) is part of the R base installation. In some scenarios, particularly for loops, significant speed gains are achieved just by adding a couple of extra lines of code. Essentially the *compiler* converts the R commands from a *sourceable* file into instructions that are executed by a virtual machine (rather loosely think in terms of Java and its runtime environment). The actual speed gains using the compiler cannot be generalized; they will be null or marginal for functions written in low-level languages and for straight R functions in packages that already make use of it. As a very rough approximation, performance improvements of around 30 % are achievable. While not dramatic, it could still mean the difference between R being a viable platform to solve a problem or having to resort to other languages instead.

The package itself has quite a few functions, here we will focus only on a single one, but further details can be found in the R help files for the package (and it is a highly recommended read).

To illustrate, as already discussed in Chap. 4, let's do something that R really does not like—a double loop (in the example we calculate allele frequencies from the matrix of genotypes). This takes about 38.7 s.

```
> test = readRDS("chapter7/genotypes.rds")
> T1=Sys.time()
> freqA=numeric(50000)
> freqB=numeric(50000)
> for (i in 1:50000)
+ {
+    hold=0
+    for (j in 1:1000) hold=hold+test[i,j]
+    freqB[i]=hold/2000
+    freqA[i]=1-freqB[i]
+ }
> print(Sys.time()-T1) # show runtime

Time difference of 38.73523 secs
```

The easiest way to use the compiler is by adding the following lines at the beginning of the code

```
> library(compiler)
> enableJIT(3)
```

The command *enableJIT* enables/disables just-in-time compilation. Use argument 0 to disable and 3 to compile loops before usage. The JIT compiler can also be enabled/disabled for all R jobs by setting the start-up environment variable R_ENABLE_JIT to a value between 0 and 3 (see details of flags in the R help files). Now, running the double loop again takes only 13.3 s, almost three times faster.

```
> T1=Sys.time()
> freqA=numeric(50000)
> freqB=numeric(50000)
> for (i in 1:50000)
+ {
+    hold=0
+    for (j in 1:1000) hold=hold+test[i,j]
+    freqB[i]=hold/2000
+    freqA[i]=1-freqB[i]
+ }
> print(Sys.time()-T1) # show runtime

Time difference of 13.29133 secs
```

7.4 Managing Memory

R was not originally designed for large datasets and, quite frankly, it is a memory *hog*. Datasets are getting larger and larger, and R not only stores everything in RAM but tends to duplicate objects unnecessarily as well. The footprint for the same operations in C can take less than a quarter of the memory used by R. The most efficient solution is to simply get a bigger machine and more memory—genomic data are expensive; is it really sensible to try to run the analysis on a $1,000.00 computer? It might be doable, but performance will take a significant hit. First, forget about 32-bit operating systems, they cannot make use of all available memory; e.g., in Windows the maximum for a single process is around 3 GB. A further limitation in 32-bit R for Windows is the size of a single object. The maximum contiguous memory allocation is around 2 GB, so if your object is larger than that, which often will be the case with genomic data, it cannot be stored as a single variable. There is no specific allocation, de-allocation and garbage collection in R, or not at least what a programmer would expect. Garbage collection is automatic but memory is not released when objects are deleted, they will only be excluded when needed. For example:

```
> test=matrix(123,10000,15000)
> memory.size()
```

[1] 1166.76

returns around 1.15 GB. After the variable is *deleted* with *rm(test)*, a call to *memory.size* still returns 1.15 GB. An explicit call to the garbage collector (with the *gc* function) will release the memory.

```
> rm(test)
> memory.size()
```

[1] 1166.8

```
> gc()
```

```
            used (Mb)  gc trigger  (Mb)    max used    (Mb)
Ncells 182094   9.8       407500  21.8      350000    18.7
Vcells 373779   2.9   126509043 965.2   150379806  1147.4
```

```
> memory.size()
```

[1] 18.08

This now returns 18M which, for this particular case, is just the footprint of R itself. Generally non-release of memory does not cause any problems because as soon as R needs the memory the garbage collector is called. If the OS requests memory the collector is also triggered, but if a very large object is removed it can be worthwhile to manually call *gc()*, even if only for the *psychological comfort* of other users on a server who might notice that your R process is eating up 200 GB of RAM.

Of more concern is that, if we create a matrix then overwrite it, R will double the memory used. For example

```
> test=matrix(123,10000,15000)
> test=matrix(321,10000,15000)
> memory.size()
```

[1] 2306.95

test is now (temporarily) using 2.3 GB, we would not be able to do this for two matrices in 32-bit R for Windows without running out of memory. But we would expect to be able to change the values of these matrices without any problems. This is not the case. Start a new R-32 session and type:

```
> test1=matrix(123,10000,15000)
> test2=matrix(321,10000,15000)
> test1=test1+1
```

Error: cannot allocate vector of size 1.1 Gb

R does not replace values, it makes a copy of the entire object with the modifications. This effectively means that to be able to manipulate the matrix you will need at least the size of the object in free memory (once finished memory is released, but it has to be available for the operation). On the positive side this only happens once and repeated changes will not blow out the memory usage any further (Chambers [16] discusses the programming logic behind this in detail). Trying to split the problem with *apply* will not work either. Now in R-64 type:

```
> test=matrix(123,10000,15000)
> test=apply(test,1,function (x) x+1)
> memory.size()
```

```
[1] 4595.96
```

This will take up 4.6 GB and is much slower to run. The last resort is a *for* loop and it will do better at 1.74 GB (saves around 450M of memory), but it is painstakingly slow. Again in a new R-64 session, type:

```
test=matrix(123,10000,15000)
for (i in 1:length(test)) test[i]=test[i]+1
memory.size()
```

```
[1] 1742.7
```

An easy workaround is to split objects into chunks, work on these one at a time and then modify the object.

```
test=matrix(123,10000,15000)
hold=test[1:1000,]+1
test[1:1000,]=hold
memory.size()
```

```
[1] 1277.07
```

This takes up around 1.3 GB—much more manageable and easy to scale between performance (number of chunks) and memory (size of chunks) given the hardware restrictions.

The limitations of working with large datasets in R are recognized, a package that aims to address this is *bigmemory*. For the example we have been using, a *bigmemory* matrix will occupy only a fraction of the memory needed by R's default matrix. If this still does not work, the package allows using disk storage as well.

```
library(bigmemory)
test=big.matrix(init=123,nrow=10000,
+ ncol=15000,type="integer")
memory.size()
```

```
[1] 25.36
```

The downside is that most R functions will not work on the *big.matrix* object (our simple *test* = *test* + 1 will return an error) and it can be rather convoluted to make use of it. Additional packages are available to facilitate usual tasks and are worth exploring (e.g., *bianalytics, bigtabulate, bigalgebra*).

One additional point about memory is that R does not copy an object immediately.

```
> test=matrix(123,10000,15000)
> test2=test
> memory.size()

[1]  1166.76
```

This will still use the same 1.15 GB of memory. But as soon as an element is changed the matrix is instantiated.

```
> test2[1,1]=124
> memory.size()

[1]  2306.88
```

This seldom has any practical effect but it can be confusing why a single change to one element of a matrix suddenly causes an out of memory error (and of course the matrix has to be copied as well, which makes an apparently small change seem very slow to execute).

7.5   Parallel Computation

Parallel computation became a hot topic with the shift from single-core to multi-core architectures, but programs and programming practices are lagging a little behind the technology; mainly because there is still more work to develop a program that runs in parallel instead of on a single thread. Currently, applications written in R are easier to parallelize than in *C* or *FORTRAN*; but this gap is closing rapidly since parallel development tools are evolving quickly. Some R packages/functions already run in parallel and are typically transparent to the user, but it is also easy to parallelize iterative routines in R (albeit this can be rather problem specific of course). In theory we could run a job eight times faster with eight processors but there is some overhead to control it all, so the speed gains are not linear and they can plateau at some point (more on this later).

Some tasks are hard to parallelize, e.g., numerical integration of systems of differential equations which depend on the previous state of all equations. On the other hand, some problems are *embarrassingly parallel*. Luckily a lot of genomic work falls in this category. For example, in a single SNP analysis each test can be performed independently from all others. Let's go back to our SNP data from Chap. 3. For simplicity we will use a file of *weights* (randomly generated as we did before) and a file with genotypes already filtered for bad SNP and in the *AA*, *AB*, *BB* and *NA* format. All we want to do is get the *anova* significance *p*-value for each SNP. The script would be

```
> genotypes=read.table("chapter7/SNPparallel.txt",
+    header=T,sep="\t")
> weight=read.table("chapter7/Weightparallel.txt",
+    header=T,sep="\t")$weight

> # function for single SNP regression
> pvalfunc=function(trait,snp)
+ {
+   if (length(levels(snp))>1)
+       anova(lm(trait~snp))[[5]][1]
+   else NA
+ }

> T1=Sys.time()
> pvals=apply(genotypes, 1,
+    function(x) pvalfunc(weight,factor(t(x))))
> pvals=data.frame(snp=row.names(genotypes),
+    pvalue=unlist(pvals))
> print(Sys.time()-T1)
```

```
Time difference of 2.930751 mins
```

Pretty much what we have already done in Chap. 3. *Sys.time* was again used to measure how long it took to run the analysis—on my machine 2.93 min.

Before discussing how to run the association analysis in parallel it is worthwhile mentioning that a really simple way of parallelizing a large task is to simply split it into smaller chunks and run each one on a different processor. This takes a bit of thinking on how to split the data and collate results, but it is very easy to do. In the folder for this chapter there are four scripts (*scp1.r, scp2.r, scp3.r* and *scp4.r*) that are exactly the same as above but each one runs a subset (15,000 in the first 3 and the rest in the fourth) of the SNP. We can simply open four R sessions and run each script in one of the sessions; then when they are all finished, read in the results to collate into a single file. This example is too small to be worth the additional effort but it is sometimes handy with large datasets. So that this exercise does not become too trivial let's use *Rcmd.exe* for this. *Rcmd.exe* is a way for R to run batch jobs in Windows (it is installed in the *bin* folder of the R path). *Rcmd* is also used to build packages for Windows. From a DOS prompt (*not* in R), navigate to the folder for Chap. 7 and then type

```
start Rcmd.exe BATCH scp1.r
start Rcmd.exe BATCH scp2.r
start Rcmd.exe BATCH scp3.r
start Rcmd.exe BATCH scp4.r
```

Fig. 7.1 Running batch jobs with *Rcmd*

as shown in Fig. 7.1. Each run took about a minute to complete (except the fourth and smaller run—20 s). We obtained around a threefold gain in time; of course here it is of little consequence since it is a small job.

Some details of the commands. The command *start* starts an independent job on a new DOS window and the OS will sort out the allocation of processors for you. If you don't use *start* the batch job will run sequentially, which beats the entire effort. With a bit of creativity it's quite simple to make better use of resources (albeit rather inelegantly). *BATCH* is the parameter for *Rcmd* to run a script. We could also run this with a *.bat* file which is just a script for running a DOS job (in the chapter's folder there is a file called *RunSNPp.bat*). It is a simple text file quite similar to what we ran in the command prompt (only difference is we gave a *name* to each window). You can run the *bat* file by typing its name on the command prompt (Fig. 7.1) or just double click on the file. Note that all this is rather Windows centric and for *Rcmd* to work it will need to be in the path (alternatively add the full path to it). The Linux equivalent is *R CMD BATCH* (this also works in Windows).

Of course R also has options for *real parallelization*. Interest in parallel computation in R has grown rapidly in the last few years. This is mainly driven by the rapid increase in the size of datasets, particularly data derived from genomic projects which are outstripping hardware performance. And also, the strong drive toward Bayesian approaches which are computationally demanding. A variety of packages and technologies have been developed for parallel computing with R. This is a very active area of research so the landscape is changing quite rapidly; a good starting point for the current state of affairs is the CRAN Task View: *High-Performance and Parallel Computing with R* (http://cran.r-project.org/). In a nutshell, there are packages to target different platforms (multicore machines, clusters, grid computing, GPUs) and these either provide low-level functionality (mainly communication layers) or easier to use wrappers that make the low-level packages more accessible for rapid development/deployment. Currently, the most widely used approach for parallel computing is the MPI (Message Passing Interface) which is supported in R through the *Rmpi* package. At a higher level the package *snow* [111] provides easy to use functions that hide details of the communication layer and allows communication using different methods (PVM,

MPI, NWS or simple sockets). An even higher level wrapper is *snowfall* which wraps the *snow* package. Herein we will illustrate how to parallelize single SNP regressions using *snowfall* and simple socket communication; a detailed overview of parallel methods/packages in R is given in Schmidberger et al. [98] and a good introductory tutorial is Eugster et al. [28]. For a more nuts and bolts view, Tierney [110] is a good read. To setup *snowfall* is quite simple:

```
>library(snowfall)

Loading required package: snow

> # set max number cpus available - default: 32
> sfSetMaxCPUs(number=128)
> sfInit(parallel=TRUE,cpus=4, type="SOCK",
+    socketHosts=rep("localhost",4))

R Version:  R version 3.1.1 (2014-07-10)

snowfall 1.84-6 initialized (using snow 0.3-13):
parallel execution on 4 CPUs.
```

The first line loads the snowfall library and then the parameter *sfSetMaxCPUs* is used to change the default maximum number of CPUs that can be accessed (in this example the value was set to 128 cores but it will only work if the machine/cluster actually has this number of nodes). If less than 32 nodes/cores will be used there's no need to change the parameter—this is simply a safeguard to avoid reckless overloading of shared resources. The function *sfInit* initializes the cluster. The parameter *cpus* stores the number of processors to be used (4 in the example). Here we are using simple socket connections (*type="SOCK"*) but MPI, PVM or NWS could be used for other cluster modes. Note that for these to work the packages *Rmpi*, *rpvm*, or *nws* would also have to be installed (and the cluster setup in these modes). Sockets do not require any additional setup but they are slower than other modes. The argument *socketHosts* is a list of cpus/computers in the cluster, for processors on the same machine *localhost* (here repeated four times—once for each node) can be used but IP addresses or network computer ids can be used for distributed systems. One point of note and common pitfall when working with clusters across different physical machines is that the socket connection is tunneled through SSH. This has to be setup and accessible to R for between computer communication to work. In Windows, firewall blocking rules may be triggered the first time the parallel nodes are initialized, make sure to allow traffic if the firewall *popup* appears.

With this the cluster is up and running. After the job has finished, the cluster can be stopped with

```
> sfStop()

Stopping cluster
```

This is all that is needed to setup a simple cluster in R. Now let's illustrate how to parallelize the single SNP regressions. All that is needed is a small change from the sequential code we used before that took around 3 min to run. To parallelize the job, after setting up the cluster, simply use the *sfApply* function which is a parallel version of *apply*. It is simply a matter of replacing *apply* with *sfApply* in the previous code for the association analysis. The full code on eight cpus is

```
> library(snowfall)
> sfInit(parallel=TRUE,cpus=8, type="SOCK",
+    socketHosts=rep("localhost",8))

snowfall 1.84-6 initialized (using snow 0.3-13):
parallel execution on 8 CPUs.

> T1=Sys.time()
> sfExport(list=list("weight","pvalfunc"))
> pvals=sfApply(genotypes,1,
        function(x) pvalfunc(weight,factor(t(x))))
> sfStop()

Stopping cluster

> print(Sys.time()-T1)

Time difference of 24.53902 secs
```

This took 24.5 s to run, around seven times faster. Runtimes do not improve linearly with the number of processors used (some of the reasons for this are detailed below) but the speed gains are still considerable, particularly with long runs.

In the parallel code an additional function was used

```
> sfExport(list=list("weight","pvalfunc"))
```

Each node in the cluster needs to have access to the data structures it has to work with. In this example the nodes had to use the phenotypic (*weight*) data to run the analyses and needed the function to calculate the *p*-values of the anova. Objects that will be used by the cluster have to be exported with *sfExport*. An important point is that objects exported for use by the cluster are duplicated on each node, this can take up sizable quantities of memory if the exported objects are large. Once the run is completed memory is released, but some thought has to be given as to whether there is enough memory on each node of the cluster (in distributed systems) or in total (for jobs parallelized on the same machine).

We allocated eight processors for the run, it took just 24.5 s to complete with the advantage of not having to use batch files and merging outputs as we did above. For details on the syntax consult the *snowfall* help files. Consider using parallelization alongside databases to improve performance in GWAS.

To conclude this section, there are a few points to consider when running R in parallel.

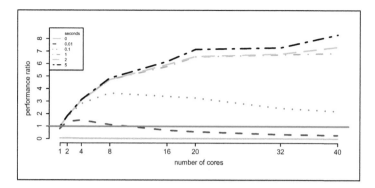

Fig. 7.2 Comparison of performance of parallel runs for varying number of nodes and various time lengths spent on independent computations within nodes. The labels on the x axis represent the actual number of cores used. The straight line through 1 (y axis) represents the sequential runtime (baseline). Figure reproduced from [44], with permission

First, computation is much faster than communication. With parallel computing it is important to minimize the transfer of data between nodes. For example, under the exact same conditions, a run takes 50.6 s and the function returns a single value (as, e.g., the above single SNP analysis). If instead, the function each time returned a vector of 10,000 numbers, the runtime increases to 91.8 s and with 100,000 numbers the runtime is 112.2 s. Considerable speed gains can be achieved by minimizing the shuffling of data between nodes.

Second, the number of calls made to the nodes also impacts performance. In practice, the more time the workers independently spend on the computation itself, the more efficient the use of the computational resources will be. Ideally the actual calculations will be the slowest aspect of the computation, and not the number of times they have to be performed. But this is not necessarily always the case and frequently computations on the nodes are trivial and the workers spend hardly any time computing, but do spend a lot of time communicating and transferring data (particularly if networks are involved). In these cases it is worthwhile to partition the problem into larger subsets but there will be some additional coding overhead.

Figure 7.2 illustrates this last point quite clearly by varying the number of workers (between 1 and 40) and computing times (0, 0.01, 0.1, 1, 2, and 5 s); i.e., the number of seconds that each node spends computing before returning results to the master. Results are shown as a ratio between sequential versus parallel runs. When the node spends 0 s on the computation (just a call to the node and returns) performance is poor due to the communication overhead incurred. As more time is spent on the computation, the gains obtained through additional cores become more evident but the compromise between computation and communication is still clear.

For times 0.01 and 0.1 s the ideal numbers of cores are respectively 4 and 8 (1.6 and 3.7 times faster). Adding more workers after this reduces the performance. For calls of 1 and 2 s the performance starts to taper off after 20 workers. Take the 1 s scenario—with 20 workers the performance is 6.6 times better than the sequential job but this only increases to 6.8 with 40 workers. Computations of 5 s show the best results in terms of speed gains but while the performance has not yet plateaued there is evidence of diminishing returns. In practice, performance is a balance between computational times and the number of cores used. Once performance starts to taper off there is little value in adding more workers. Either performance will decline due to additional communication overhead or there will be unnecessary strain placed on computational resources, which becomes quite relevant in shared environments.

Third, performance gains through parallelization are not linear (i.e., four cores do not perform four times better than one core) but rather incremental. For example, consider the 5 s computation time in Fig. 7.2—with 40 cores the performance is 8.3 times better (far from a 40-fold performance boost). While this may not sound optimal, if actual runtimes are taken into consideration the picture becomes much more interesting: the sequential run takes 36.7 min whilst the 40 core run takes 4.4 min. Returning to the example we have been working with, the sequential run took 175.8 s to run and parallel runs with 2, 4, 8 and 16 cores took respectively 88.08, 45.74, 24.52, and 19.67 s. The speed fold gains were 1.99, 3.84, 7.17, and 8.93—good improvements up to 8 cores but with 16 nodes performance starts to taper off.

Last point, there is a fixed time lag to initialize the communication protocols which is around 1.5 s. That is why the parallel run on a single worker performs worse than the sequential one (Fig. 7.2).

In summary, the key points to consider when setting up parallel tasks in R are:

- the more time spent on computations, the more useful it becomes to parallelize the process and more cores can be used,
- communication is more expensive than computation, try to minimize data transfer and calls to nodes whilst maximizing computation within workers,
- performance gains are not linear to the number of cores and can reach a *saturation* point; but actual computing times can be significantly lower.

7.6   External Interfaces in R

A lot can be done in R, but sometimes we have code or other applications that we need or want to use. R can be dynamically linked to compiled code in *C* or *FORTRAN* (and also other languages to various degrees); this opens the possibility of using prior code or to develop code especially designed to solve a computationally intensive task. Results can then be sent back into R for further downstream analyses. A large part of R is written in *C*, *C++* and *FORTRAN*, so it is relatively straightforward to compile, link, call, and execute routines written in these languages. Other languages are also supported to some extent or other (see packages *SJava*, *RSPerl*, and *RSPython*).

We will show a simple example of how to run a *C++* routine in R at the end of this section but, to start; the easiest way to link to external routines is to simply run applications compiled in another language straight from within R.

All data preprocessing and setup of files could be done in R, then call the program from within R and finally read the results back into R for further analyses. The R function for this is *system*. By default *system* will wait for the job to finish before continuing, but you can also work asynchronously using the argument *wait=FALSE* or redirect the output to R (provided it is a console program) with *show.output.on.console=TRUE* and *intern=TRUE*. Program arguments can also be passed using *system* but note however that command line arguments cannot be passed using *system* in Windows. Use *shell* instead or the newer implementation *system2* (see the R help files for details). To illustrate let's use *Notepad* to look at the phenotypes in the file *Weightparallel.txt*.

```
> system("notepad.exe chapter7/Weightparallel.txt",
        wait=F,invisible=F)
```

This opens the file in *Notepad*. In Linux or MacOS replace *Notepad* by another text file editor (e.g., *gedit* in Linux). In Windows, to see programs that have a graphical interface the argument *invisible=F* has to be used; not necessary for Linux or MacOS. Now, for something more useful, the *DECIPHER* package has the *AlignSeqs* function to perform multiple sequence alignments, but *ClustalW* is the most widely used programs for this common bioinformatics task. To execute the code below (this will take quite some time to run) *ClustalW* needs to be installed in the directory of the chapter (or elsewhere, but change the path accordingly). *ClustalW* and the newer *Clustal Omega* are freely available from http://www.clustal.org/. Note that the Clustal parameters have to be on the same line and are space separated—below they are shown on separate lines for clarity purposes only.

```
> # clustal W - for DNA
> system(paste("chapter7/clustalw2
    -INFILE=chapter7/sequenceData.txt
                -OUTFILE=chapter7/aligned.fasta
                -OUTPUT=FASTA
                -TYPE=DNA
                -PWDNAMATRIX=IUB
                -PWGAPOPEN=10
                -PWGAPEXT=0.1
                -ALIGN
                -DNAMATRIX=IUB
                -GAPOPEN=10
                -GAPEXT=0.2
                -GAPDIST=5"))

CLUSTAL 2.1 Multiple Sequence Alignments

Sequence type explicitly set to DNA
Sequence format is Pearson
```

```
Sequence  1:  HamadryasBaboon        16521 bp
Sequence  2:  BarbaryMacaque         16586 bp
Sequence  3:  RhesusMacaque          16564 bp
Sequence  4:  LarGibbon              16472 bp
Sequence  5:  BorneanOrangutan       16389 bp
Sequence  6:  WesternGorilla         16364 bp
Sequence  7:  Bonobo                 16563 bp
Sequence  8:  CommonChimpanzee       16554 bp
Sequence  9:  WesternChimpanzee      16561 bp
Sequence 10:  Neanderthal            16565 bp
Sequence 11:  HumanCambridge         16569 bp
Sequence 12:  HumanYoruba            16571 bp
Sequence 13:  WesternLowlandGorilla  16412 bp
Sequence 14:  BoaConstrictor         18905 bp
Sequence 15:  Crocodile              17900 bp
Sequence 16:  Dog                    16727 bp
Sequence 17:  Wolf                   16729 bp
Sequence 18:  Buffalo                16359 bp
Sequence 19:  BosIndicus             16341 bp
Sequence 20:  BosTaurus              16338 bp
Sequence 21:  Auroch                 16338 bp


Start of Pairwise alignments
Aligning...
Sequences (1:2) Aligned.  Score:  85
Sequences (1:3) Aligned.  Score:  82
Sequences (1:4) Aligned.  Score:  79
Sequences (1:5) Aligned.  Score:  79
Sequences (1:6) Aligned.  Score:  79
Sequences (1:7) Aligned.  Score:  80
Sequences (1:8) Aligned.  Score:  79
Sequences (1:9) Aligned.  Score:  79
Sequences (1:10) Aligned.  Score:  77
...
Sequences (17:20) Aligned.  Score:  76
Sequences (17:21) Aligned.  Score:  77
Sequences (18:19) Aligned.  Score:  88
Sequences (18:20) Aligned.  Score:  87
Sequences (18:21) Aligned.  Score:  88
Sequences (19:20) Aligned.  Score:  98
Sequences (19:21) Aligned.  Score:  98
Sequences (20:21) Aligned.  Score:  99
Guide tree file created: [chapter7/aligned.dnd]
There are 20 groups
```

```
Start of Multiple Alignment
Aligning...
Group 1: Sequences:    2        Score:286459
Group 2: Sequences:    3        Score:281562
Group 3: Sequences:    2        Score:313499
Group 4: Sequences:    3        Score:307141
Group 5: Sequences:    2        Score:314341
Group 6: Sequences:    3        Score:312601
Group 7: Sequences:    6        Score:289036
Group 8: Sequences:    2        Score:309608
Group 9: Sequences:    8        Score:287535
Group 10: Sequences:   9         Score:278283
Group 11: Sequences:   10        Score:275077
Group 12: Sequences:   13        Score:258438
Group 13: Sequences:    2        Score:316773
Group 14: Sequences:    2        Score:309583
Group 15: Sequences:    3        Score:307827
Group 16: Sequences:    4        Score:287818
Group 17: Sequences:    6        Score:252796
Group 18: Sequences:   19        Score:233955
Group 19: Sequences:   20        Score:213287
Group 20: Sequences:   21        Score:206131
Alignment Score 17776121
firstres = 1 lastres = 21273
FASTA file created!
```

```
Fasta-Alignment file created [chapter7/aligned.fasta]
```

The file *sequenceData.txt* contains mitochondrial DNA sequence data for various species downloaded from NCBI. We performed the multiple sequence alignment in ClustalW (check the program's documentation for information of the parameters used) and can now, for example, use it to build a phylogenetic tree as shown in Fig. 7.3. This is quite easy to do using the package ape.

```
> library(ape)
> difmat=read.dna("chapter7/aligned.fasta",
+   format="fasta")
> difmat=dist.dna(difmat,model="raw")
> difmat=hclust(difmat,"average") # simple UPGMA
> plot(as.phylo(difmat))
> axisPhylo()
```

This is just a simple distance tree to illustrate how R and other programs can be merged together into an analysis pipeline (it is handy for, e.g., the RNA-seq analysis in Chap. 5). Readers interested in phylogenetic work might want to start with the excellent book of Paradis [84] and the *ape* package.

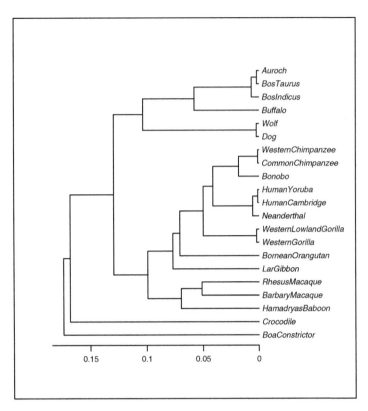

Fig. 7.3 Example of a simple *UPGMA* phylogenetic tree built with the *ape* package using mitochondrial sequence data aligned in *ClustaW*

Data used in the example was downloaded within R from NCBI using the accession ID (*refseq*) of the mitochondrial sequence for each species. For example, for the cattle and human data we can use the following code to query NCBI using the *Entrez Programming Utilities* (for details see *eutils.ncbi.nlm.nih.gov*):

```
> # get mitochondrion sequences
> # e.g. Bos taurus and human
> # accession IDs
> mitochondria=c("NC_006853.1","NC_012920.1")
> for (i in 1:length(mitochondria))
+ {
+   link=paste("http://eutils.ncbi.nlm.nih.gov/entrez/
+           eutils/efetch.fcgi?db=protein&id=",
+           mitochondria[i],
+           "&rettype=fasta&retmode=text",sep="")
+
+   download.file(link,
```

```
+        destfile=paste("chapter7/MT",i,".txt",sep=""),
+        mode = "w")
+ }
```

```
trying URL 'http://eutils.ncbi.nlm.nih.gov/entrez/
   eutils/efetch.fcgi?db=protein&id=NC_006853.1
   &rettype=fasta&retmode=text'
Content type 'application/octet-stream' length unknown
opened URL
downloaded 16 Kb
```

```
trying URL 'http://eutils.ncbi.nlm.nih.gov/entrez/
   eutils/efetch.fcgi?db=protein&id=NC_012920.1
         &rettype=fasta&retmode=text'
Content type 'application/octet-stream' length unknown
opened URL
downloaded 16 Kb
```

7.6.1 Linking R to C++

At some point you may need to use code written in another language in R. We will illustrate here with a simple example using *Rcpp*. Readers interested in the topic could start with Chap. 6 of Gentleman's *R Programming for Bioinformatics* [39] and then move on to Chamber's [16] book. Rizzo's [94] text while it does not discuss language interfacing might also be appealing due to its more algorithmic approach. Good entry points for *C++* programming are [88] (accessible pure *C++* introductory text) and [27] (a blend of *C++* and R, makes it easier to work with *C++* for those who are already familiar with R). The best reference for those who already know *C++* is the applied text *Seamless R and* C++ *Integration with Rcpp* [26] written by Eddelbuettel, the author of the package.

Rcpp is fully described in [26] and a must read for programmers making extensive use of *C++* in R. In simple terms, *Rcpp* acts as the bridge between *C++* and R, facilitating communication between the languages. On the *C++* side of things it is just a couple of extra lines: we need to add a new header (*Rcpp.h*) and a *using* statement for the namespace:

```
# include <Rcpp.h>
using namespace Rcpp;
```

Functions that will be *exposed* to R have to start with

```
// [[Rcpp::export]]
```

To illustrate, the code below receives two numeric values from R, adds them up and returns the result to R (this is the full *C++* code).

```
# include <Rcpp.h>
using namespace Rcpp;

// [[Rcpp::export]]
int simpleSum(const int x, const int y)
{
  return x + y;
}
```

On the R side it can also be quite simple. First download, install, and load the *Rcpp* package (will also need additional compiler tools installed). Then the *C++* code can be compiled *on the fly* (using *sourceCpp*) and the functions exported with (*Rcpp::export*) will become available for the session (here only *simpleSum*). Once *sourced* all functions become available for use, similar to other internal functions in R.

```
> library(Rcpp)
> sourceCpp("chapter7/Cpp/SimpleSum.cpp")

> # function: simple sum of two numbers
> simpleSum(5,3)
```

```
[1] 8
```

The function *sourceCpp* compiles a *dll* (link library) in memory and makes it available for a single R session—the library is deleted when the session is closed (but it will be saved together with the workspace). This is probably the easiest way to interface with *C++* from R. But there are a few caveats: the library will have to be compiled and linked each time you want to use the *C++* functions; depending on the number of functions, it can take some time to compile. More of a problem is the lack of portability if you want to share the library with others. Most Linux installations will have the necessary tools to compile the library but Windows does not. There are quite a few compatibility problems with different compilers, the easiest way to get things to work is to install the *Rtools* (includes the *gcc* compiler and various other applications needed to build R packages for Windows from source). *Rtools* can be downloaded from *cran.r-project.org/bin/windows/Rtools*. Make sure to edit the path so that Windows knows where to find the tools. In summary you will need the *Rcpp* package and *Rtools* (installed and visible in the path) for this to work.

When using *C++*, or other languages for that matter, pay attention to error handling. The *C++* code above is *unprotected* and could result in R *hanging* on you.

A less *ephemeral* solution for the library is to compile and save it to disk. This is rather tricky to do manually but we can use the functionalities for building packages to make this easier. The *Rcpp.package.skeleton* function will create all that is needed to compile the library. First set the working directory to the folder with the *SimpleSum.cpp* file—the function does not seem to be able to resolve partial paths (preferably no spaces in the path either); then create the package structure.

```
> setwd("chapter7/Cpp")
> Rcpp.package.skeleton("egDll", example_code = FALSE,
+    cpp_files = c("SimpleSum.cpp"))

Creating directories ...
Creating DESCRIPTION ...
Creating NAMESPACE ...
Creating Read-and-delete-me ...
Saving functions and data ...
Making help files ...
Done.
Further steps described in './egDll/Read-and-delete-me'.

Adding Rcpp settings
 >> added Imports: Rcpp
 >> added LinkingTo: Rcpp
 >> added useDynLib directive to NAMESPACE
 >> added importFrom(Rcpp, evalCpp)
                          directive to NAMESPACE
 >> copied SimpleSum.cpp to src directory
```

This will create a new directory called *egDll* with the basic folders and files needed to build an R package. The argument *cpp_files* are the names of the source code files (or just one, as here). The name of the package (here *egDll*) will become the name of the library. Note: the steps to build an R package are quite similar to this, for details on how to make your own packages read the *Writing R Extensions* document [in the *R Console*, click on *Help* and then on *Manuals* (*in PDF*)]. We can now compile the *package* with

```
> system ("R CMD INSTALL --build egDll")
```

This will create a file called *egDll_1.0.zip* with 32- and 64-bit versions of the DLL inside the *libs* folder. Now extract the DLL into the Cpp folder and we are ready to load it and use it in R. But before directly accessing the library, note that in the process we created a *real* R package that was installed in the *library* folder of the R installation path and the *egDll_1.0.zip* is the same as any other R library from CRAN or Bioconductor. We could simply load the library and access the functions in the same way as with any other package (do this in a new *clean* R session):

```
> # using as a library
> library(egDll)
> simpleSum(6,9)

[1] 15
```

It is rather unpolished but this is all that is needed for an *in-house* package. To link directly to the library (remember to extract the DLL from the zip file and place it in the Cpp folder) we use *dyn.load* to load the library and calls to functions in the DLL are made with *.Call*:

```
> library(Rcpp)
> dyn.load("egDll.dll") # load dll
> .Call("egDll_simpleSum",5,10) # call functions
```

[1] 15

The syntax for call is *nameDLL_functionName* followed by the parameters of the function. Note that *Rcpp* has to be loaded prior to loading the DLL or using functions.

This is a bit complicated and for illustration purposes only, but ultimately all we really need are the definitions for the functions that we want to export to R. If you look in the *src* folder that was created by the *Rcpp.package.skeleton* function there will be a new file called *RcppExports.cpp* that *translates* the *simpleSum* function into a structure that can be read by R. This file looks as below.

```
// This file was generated by
// Rcpp::compileAttributes

// Generator token:
// 10BE3573-1514-4C36-9D1C-5A225CD40393

#include <Rcpp.h>
using namespace Rcpp;

// simpleSum
int simpleSum(const int x, const int y);
RcppExport SEXP egDll_simpleSum(SEXP xSEXP, SEXP ySEXP)
{
 BEGIN_RCPP
    SEXP __sexp_result;
    {
     Rcpp::RNGScope __rngScope;
     Rcpp::traits::
input_parameter<const int>::type x(xSEXP);

        Rcpp::traits::
input_parameter<const int>::type y(ySEXP);

        int __result = simpleSum(x, y);
        PROTECT(__sexp_result = Rcpp::wrap(__result));
    }
    UNPROTECT(1);
    return __sexp_result;
  END_RCPP
}
```

It is rather cryptic unless you know *Rcpp* well and it is simpler to let the function *Rcpp.package.skeleton* do the translation for us. But it you want to bypass the package building step this file can be used to compile the DLL directly with g++ and then use *R CMD SHLIB* to make a DLL to use with *dynload*.

These are just the basics to get started with *blending* C++ and R. There are many other useful features in Rcpp and they are well described in [26]. To finish this section, there are a few *Rcpp* extensions that make it easy to transfer data from R into *C++*. These are *NumericVector, NumericMatrix, CharacterVector,* and *CharacterMatrix* (these will not work with a *data.frame*, first convert to matrix and then send to *C++*). An example illustrating how to use them is shown in the *RcppExample.cpp* file (in the Cpp folder of this chapter). One of the functions calculates the row sums of a numeric matrix (*rowSumsC*). To get the row sums for the genotypes in the *RDS* file we made before use:

```
> library(Rcpp)
> sourceCpp("chapter7/Cpp/RcppExample.cpp")
> geno=readRDS("chapter7/genotypes.rds")
> sums=rowSumsC(geno)
> head(sums)

[1]    956 1006 1002   984   989 1003
```

This should be the same as using the *rowSums* function in R.

7.7 Using R Inside Other Applications

R is extremely powerful, but it is not very different from a programming language—which can make it quite complex to use. We can hide this complexity from the end user by wrapping R in a user friendly point-and-click graphical application. A nice example is *affylmGUI*, a graphical user interface for *limma*. Another example is *AffyPipe* for analysis of Affymetrix microarray data. The program is written in C# and allows the user to import the array files, set up the experiments (contrasts) of interest and run the analysis. The program will run a full analysis with quality control preprocessing, differential expression, gene ontology, and pathway analysis and create an *HTML* report, part of which is shown in Fig. 7.8. Figure 7.4 shows a screenshot of the program.

How to interact with R will depend on the programming language used. The general steps are to use (in Windows) *Rterm.exe*—R in terminal/console mode. From the GUI start a new instance of *Rterm.exe* as an asynchronous process, hide the console and redirect the *standard input, output* and *error* as streams to your application. And then it's just a matter of creating, e.g., buttons in your application which when clicked will send commands (i.e., a simple text stream with R's syntax) to R and when R returns output to the *output stream* you can handle it by, e.g.,

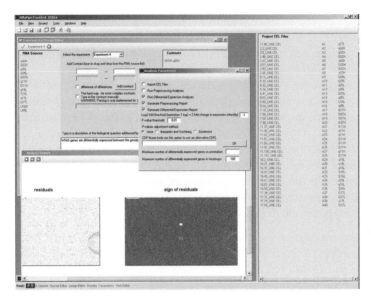

Fig. 7.4 Screenshot of *AffyPipe*, a C# frontend of an R-based pipeline for analysis of Affymetrix microarray data

building a table or making a plot. A nice side effect is that computationally intensive tasks can run remotely on a large machine while the GUI sits on the user's desktop controlling the application.

In the folder for this chapter there is a very simple example in C# (the code, solution, and executable are in the *RwithGUI* folder). If you are familiar with *.Net* programming languages it should be easy to port to other languages (e.g., Visual Basic) or adapt the project to your needs. We will not discuss the syntax here, but the program basically just does what was mentioned above: connects to *Rterm* and redirects input/output to/from the program. The GUI is shown in Fig. 7.5 and all it does is the four basic operations (addition, subtraction, multiplication and division) and it also has a line to input R code and visualize the output (shown in the black window).

Ultimately this is just the flip side of what we did before using *system* in R to execute other programs—now we used another program to run R. Apart from being useful to create simpler and more user-friendly tools for specific tasks with R as the primary engine; it can also be used to simplify programming jobs by hooking into R to access specific functions. In loose terms, R and all its packages and functions become a vast programming library that can be used in your applications.

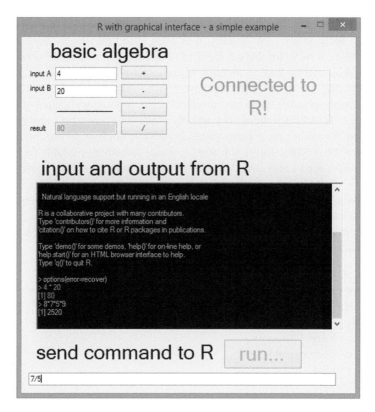

Fig. 7.5 Screenshot of a C# frontend for R using *Rterm.exe*. The code is in the folder for this chapter under *RwithGUI*

7.8   Reporting in R

Once the analyses are finished, results must be collated and presented in a *human digestible* format. We have already seen how to save outputs (tables) and plots, but it's also handy to pull it all together as part of the analysis—this ensures that the whole analysis is repeatable and, in practice, it is common to have to go back to an earlier stage and try a new parameter or test a different model or have to analyze a new batch of data of exactly the same type. It is time consuming (and irritating) to cut and paste tables and plots into a word processor for *the final result* only to find out that there was this other important covariate that somebody forgot to mention and you have to start again...

There are two goods ways of reporting in R. One is using *Sweave* which generated what you have been reading so far. *Sweave* allows mixing R code with Latex code in the same file which you can run in R to produce *tex* files for Latex. This amounts to an (executable) report which is fully reproducible (there is a strong drive in the research community to improve documentation of workflows).

If you are not familiar with Latex, it will take some time to get used to the syntax and basic setup. Conceptually, it is a tag based language—think in terms of an *HTML* equivalent for text editing. Error handling is probably the weakest part of Latex and it can be hard to identify where exactly the problems are; it is also rather unfriendly for tables (you may have noticed that they are rather wanting in this text). On the positive side, Latex produces really professional output (and tables also look good) and for formulas it's actually easier than any other word processor I have used. You don't need Latex installed on your machine to use *Sweave*, you can still get the *tex* files. But if you want to make something readable (normally a *pdf* file) you'll need Latex installed. *MiKTeX* is a good option (http://miktex.org/) and there is also a portable version that does not require a full installation. To get started have a look at the *The Not So Short Introduction to LaTeX2e* by Oetiker et al. available at http://tobi.oetiker.ch/lshort/lshort.pdf. But let's get back to *Sweave*. The only real difference between a normal script and a sweave script is the separation between latex and R. A simple example follows.

```
\documentclass[10pt,a4paper]{article}
\title{Example of an R Report using \emph{Sweave}}
\author{John Doe}
\date{}
\usepackage{Sweave}
\begin{document}
\maketitle

This report shows the density plot for
10000 numbers randomly sampled from a
normal distribution with mean = 100 and
standard deviation = 10.
 <<>>=
 rnum=rnorm(1000,mean=100,sd=10)
 summary(rnum)
 plot(density(rnum),col="blue",main="Density plot")
 dev.print(file="sweaveplot.pdf",
  device=pdf,width=8,height=8)
 @
\begin{figure}[h]
\centering
\includegraphics[width=0.8\textwidth]{sweaveplot.pdf}
\caption{Density plot of random numbers
  sampled from a normal distribution.}
\label{sweaveplot}
\end{figure}
\end{document}
```

Save this script as for example *MySweave.snw* and in R, instead of the usual *source*, use *Sweave(MySweave.snw)* this will produce a *MySweave.tex* file ready for use.

```
> Sweave("chapter7/MySweave.snw")

Writing to file MySweave.tex
Processing code chunks with options ...
 1 : echo keep.source term verbatim

You can now run (pdf)latex on MySweave.tex
```

R commands are interwoven with Latex using the symbol <<>>= to start the R code and @ at the end. If Latex is installed and in the path, we can make a *pdf* from the *tex* file using *system*

```
> system(paste("pdflatex.exe","MySweave.tex"))
```

Which will create the *pdf* (Fig. 7.6). Another option is to use the function *texi2dvi* from the package *tools*. Another useful package for working with Latex is *maDB*, it will even make tables for you. *Hint*: a common problem is that Latex tries to find a file called *Sweave.sty*, make sure it's in one of the path folders that the OS knows about (the file is usually in the R installation folder under *share/textmf*). For convenience, it is copied in the chapter's folder directory. Also note that the *MySweave.tex* file was created in the working directory and not in the folder for Chap. 7. To create the pdf using, e.g., portable MikTex, make sure both the *tex* file and the *Sweave.sty* file are in the same directory.

A second way of reporting results is in *HTML*. Even though links can be used in *pdf* files it is a bit cumbersome. *HTML* is ideal for large reports with many cross references and dynamic links to external resources. There are many packages for working with *HTML* in R. The package *annotate* can be used with function *htmlpage* to build tables with genes and links to external databases (e.g., *Entrez, GO,...*) or *pmAbst2HTML* that will build a web page with abstracts and links to *PubMed*. For Affymetrix probe ids the function *probes2table* in the *affycoretools* package will also generate an annotated *HTML* table. There are many other platform-specific packages to perform similar tasks (have a look in Bioconductor if there's an *off the shelf* solution that works for you). On the other extreme if you are comfortable with *HTML* you can just write the code straight into your script and save in a file using *write*. In between the two extremes there's the package *R2HTML* which has many functions to simplify *HTML* reporting. For example

```
> library("R2HTML")
> HTML("<h1><center>Example of an HTML report with a
+   table and plot</center></h1>",
+   "chapter7/example.html",append=FALSE)
> data=data.frame(id=paste("id",1:20,sep=""),
+   trait=c(rnorm(10,10,1),rnorm(10,12,1)),
```

Example of an R Report using *Sweave*

John Doe

This report shows the density plot for 10000 numbers randomly sampled from a normal distribution with mean = 100 and standard deviation = 10.

```
> rnum=rnorm(1000,mean=100,sd=10)
> summary(rnum)

   Min. 1st Qu.  Median    Mean 3rd Qu.    Max.
  66.78   93.20  100.10  100.20  106.80  136.30

> plot(density(rnum),col="blue",main="Density plot")
> dev.print(file="images\\sweaveplot.pdf",device=pdf,width=8,height=8)

windows
      2
```

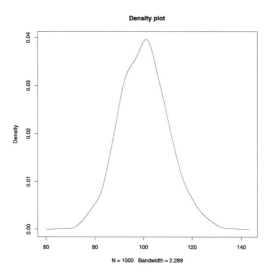

Figure 1: Density plot of sampled random numbers.

Fig. 7.6 Sample *Sweave* report

```
+    treat=c(rep("A",10),rep("B",10)))
> HTML(data,file="chapter7/example.html",
+    append=T,row.names=FALSE,Border=3,
+    innerBorder=1,align="center")
> boxplot(data$trait~data$treat,col=c("blue","red"),
+    main="Boxplot of trait by treatment")
> dev.print(file="chapter7/htmlboxplot.png",
+    device=png,width=400,height=400)
```

```
windows
      2

> HTMLInsertGraph(GraphFileName="htmlboxplot.png",
+    GraphBorder=1, Align="center", WidthHTML=400,
+    HeightHTML=400, file="chapter7/example.html",
+    append=TRUE)

[1] TRUE
```

The report is shown in Fig. 7.7. Note that *HTML* works for simple text and also dataframes or matrices. The function *HTMLInsertGraph* was used to add the boxplot—just make sure that the size of the figure are the same when you save it and in the *HTML* report or it will look pretty ugly. A real example from a microarray report generated with the program *AffyPipe* (mentioned above) is shown in Fig. 7.8.

7.9 Summary

Following the framework of this primer, we overviewed in rather broad terms some other ways of working with R and how to be more efficient when working with large datasets. If computational requirements are becoming unwieldy, it is worthwhile investigating faster ways of handling and/or analyzing the data—there is a good chance there will be a different way of coding or a function implemented in a package that can make the problem more tractable. Better ways of reading in the data and parallelization will handle most of the common bottlenecks in genomic analysis. If nothing else works, lower level languages can be used directly in R. This means that R can be just as fast as *C++* or *FORTRAN* but the advantages of rapid coding by simply using available functions will be lost. Pipelines that require many different programs can be built using R; this is particularly useful for sequence data. It is straightforward to integrate reporting and analysis into a single process in R using *Sweave* and *HTML*. Genomic data is large and involves many intermediate steps between raw data and final results—the more all these steps are automated and reproducible, the easier it will be to rerun or tweak an analysis and to avoid irreproducible human errors. Ideally, every single analysis (and dataset) should have an executable and well-documented workflow (for us, an R script). *Hint*: when working with scripts, comment and justify any *judgment calls* that had to be made and then, *programmatically* implement those decisions (e.g., if outliers are removed based on visual inspection, add the removal of the data into the script itself).

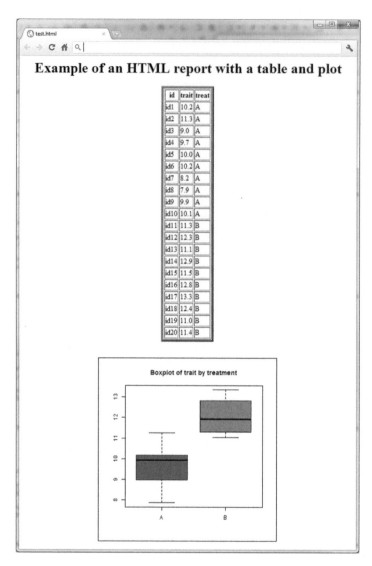

Fig. 7.7 Sample *HTML* report

7.10   Useful R Books and Packages

- *Data Manipulation in R* [105]. Lots of useful tips for working and manipulating data, but not centered on large datasets.
- *R Programming for Bioinformatics* [39]. Gentleman's book has a good overview of how to use R with other languages. The chapter on external annotation

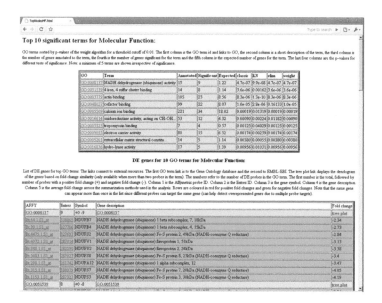

Fig. 7.8 Screenshot of *HTML* report generated by *AffyPipe*

databases and how to use them in R is a highlight. There is also some interesting material on how to work with sequence data.

- *Software for Data Analysis* [16] and *Scientific Programming and Simulation Using R* [60]. Both have a more *programmatic* style of writing—useful for readers interested in a more *under the hood* and broad understanding of R.
- *Parallel R* [74]. An objective, hands-on overview of parallelization in R with a good overview of the *snow* package for parallel computation.
- *Seamless R and C++ Integration with Rcpp* [26]. A comprehensive overview of how to integrate R and *C++* using the *Rcpp* package. And a good complement, more algorithmically focussed is *Statistical Computing in C++ and R* [27].
- *Building Bioinformatics Solutions with Perl, R and MySQL* [11]. Good introduction to the three platforms with an integrative focus for handling bioinformatics tasks.
- *CRAN Task Views*. R packages organized into themes; extremely useful with topics such as *reproducible research* and *high-performance computing* (*cran.r-project.org/web/views*).
- *Revolution R*; R with additional features for faster data handling—runtime differences are very significant with matrix manipulations (*revolutionanalytics. com*).

Chapter 8
Final Comments

And hence we come to an end of our *Primer to Analysis of Genomic Data Using R*. We have covered quite a broad range of topics—I have endeavored to keep this discussion very case oriented with the barest minimum of theory I thought I could get away with (and rather informal, as you may have noticed). I especially tried to highlight the main pitfalls that new users will come across when using R, particularly if your background is more toward biology or computer science (for statisticians R is probably rather intuitive and almost second nature). Many examples in the early chapters were purposefully inadequate in terms of the data structure which forced us to understand what R was doing with it and taught us how to handle it. But once we get used to some of these *quirks*, R is quite easy to work with and, thanks to all the contributed packages, the development time can be minimal.

Another topic discussed quite extensively was data preprocessing. With genomic data it is critical to ensure good quality data—bad data can yield very unreliable results which makes data preprocessing paramount for any genomic work. We used a rather bad microarray dataset to illustrate these issues in Chap. 5. If due consideration is not given to this step, no matter how fancy or complicated the actual analyses, the outcomes will be poor. The message is clear: invest whatever time is needed in the preliminary steps to understand the platform and the data.

By now you must have a fair idea of the wide scope of functionality available in R. The number of R books is growing steadily and the number of packages is borderline overwhelming. We only had a small taste of what is out there but I hope it was enough to make you feel comfortable using R and provided the stepping stones for building your own applications and analyzing your data. We will finish off with my shot at guessing what the future holds in store for R and genomics. But until all this comes true—most of which I'll probably get all wrong—happy *Ring*...

© Springer International Publishing Switzerland 2015
C. Gondro, *Primer to Analysis of Genomic Data Using R*, Use R!,
DOI 10.1007/978-3-319-14475-7_8

8.1   The Future: Polishing the Crystal Ball

Some futurology. For R, I expect there will be some heavy development of packages for handling and analyzing sequence data, it is a very active topic right now and we should see packages with a similar level of maturity as those available for, e.g., microarrays in the near future. Parallelization, virtualization, and high performance computing will continue to be on the agenda and will become more embedded in R itself, requiring less *conscious* coding by the users. We should see improvements in data handling and memory management to better suit current datasets. After all, datasets are becoming large (and that is an understatement) and ways of coping with dimensionality problems are a hot topic—and not just for R.

Where is all the genomic work going? In terms of GWAS we can expect denser and denser marker coverage. It's not unreasonable to speculate that all common human variants will be assayed (just the SNP would be around 11 million). Thus there is a clear trend toward direct causal association studies while LD and haplotypes become less relevant, but only in the context of association studies, of course. Livestock will lag behind a couple of years as it tends to do and will probably never get to see the same density as human SNP arrays, but will instead jump straight onto the next big wave—full sequencing. Over the next 5–10 years full (phased) sequencing should be economically viable for large scale projects. In the meantime there will be a lot of research into imputation: sequence a relatively small number of individuals and impute up to sequence level datasets genotyped at lower densities.

This of course will bring on another onslaught of analyses methods and really interesting problems of how to handle the data—a project with 10,000 samples would generate something along the lines of 30 trillion data points. For the here and now, interactions will be (in truth they already are) a hot topic—*gene × gene* interactions, networks, *gene × environment* interactions. Lots of interesting opportunities for research in this area. We will see a lot of work in merging GWAS results with functional knowledge/studies—either as supporting evidence for associations or to help weed out the noise (false positives).

Microarrays are on the verge of being replaced by next-generation sequencing. I expect the number of RNA-seq publications, both methods and applications, will continue to accelerate for the next few years. Current arrays might still have a place in biomedical applications—probably move out of the labs and into the clinics, but prices are becoming uncompetitive which could make them obsolete.

All this should happen in a rather short time frame—not good for me since I'll be proven wrong pretty soon! But the point is, the changes will be/are being dramatic and fast, and the development/availability of tools to handle the data will just become more and more important. R will continue to play a key role in this field.

Appendix A
Example QC Report for GWAS Data

Over the following pages a fully automated quality control report for SNP data is illustrated. QC metrics are the same as those discussed in Chap. 3. QC results are then mingled with standard text blurbs using *Sweave* as discussed in Chap. 7 to generate a PDF report. This approach allows producing fully reproducible analyses and automates reporting in a format that is *human digestible*. The report was generated by the program *snpQC* [45].

© Springer International Publishing Switzerland 2015

C. Gondro, *Primer to Analysis of Genomic Data Using R*, Use R!,

DOI 10.1007/978-3-319-14475-7

Quality control report for Example 50K simulated SNP chip data

John Doe and Jane Doe

May 30, 2013

Abstract

This report encompasses the quality control summary for the Example 50K simulated SNP chip data. A total of 83 samples were genotyped for 54977 SNPs. Quality control was performed across samples, across snps and on physical location. The results for each of these and the filtering criteria used are discussed herein.

1   QC filtering results

Out of the 83 samples, 2 did not pass the filtering criteria (2.41%). From the 54977 SNPs 4757 were excluded (8.65%). Out of the total 4563091 genotypes, 502210 were excluded (11.01%). Filtering criteria consisted of QC metrics across SNPs, across arrays and on the physical mapping as detailed in the following sections.

Table 1 summarizes the number of SNPs and samples rejected for each QC criterion. Note that many of these overlap across criteria, thus the final numbers are not simply a sum of the rejection numbers for each criterion.

The correlation criterion for samples was not used to reject samples but simply to flag potential replicates which should be checked before further analyses. Correlation includes SNPs and samples flagged as bad which makes samples less similar than they should be. The correlation matrix should be used only for QC purposes. For downstream analysis the GRM constructed after data filtering should be used.

2   SNP statistics

In this section the descriptive statistics for the dataset on a per SNP basis are discussed. Figures 1 and 2 illustrate the difference between good and bad quality genotypes.

2.1   SNP call rates

The number of SNPs with a call rate higher than 99.5% was 74.3% (Table 2 and Figure 3). As a rule of thumb around 90% of the snps would be expected to have a call rate above 99.5% and less than 2% would have call rates under 90%. In some cases the bulk of the data may be just below, in the 0.99-0.995

Table 1: Summary of SNPs and samples rejected for each QC criterion.

SNP criteria	number
>5 percent genotyping fail	1432
median GC scores <0.5	2078
all GC scores 0	650
GC <0.5 in less than 90 percent samples	2557
100 percent homozygous	178
MAF <0.01	129
heterozygosity 3SD	6
Hardy-Weinberg at 1e-15	131
sample criteria	number
call rates <0.9	2
correlation >0.98	0
heterozygosity 3SD	0
mapping criteria	number
Chromosome 0	317
Chromosome X	1502
Chromosome Y	66

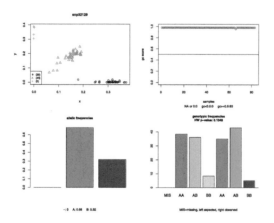

Figure 1: Example of a good quality SNP. Top left: clustering for each genotype (non calls are shown as black circles). Top right: GC scores. Bottom left: non-calls and allelic frequencies (actual counts are shown under the histogram). Bottom right: genotypic counts, on the left hand side the expected counts and on the right the observed counts; the last block shows number of non-calls.

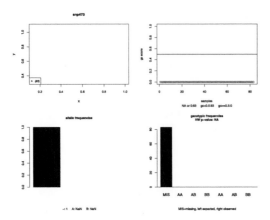

Figure 2: Example of a bad quality SNP. Top left: clustering for each genotype (non calls are shown as black circles - here all samples). Top right: GC scores. Bottom left: non-calls and allelic frequencies (actual counts are shown under the histogram). Bottom right: genotypic counts, on the left hand side the expected counts and on the right the observed counts; the last block shows number of non-calls.

3

Table 2: Call rates for SNPs.

rate	count	frequency
<0.9	2557	0.047
0.9-0.95	568	0.010
0.95-0.99	10988	0.200
0.99-0.995	0	0.000
>=0.995	40864	0.743

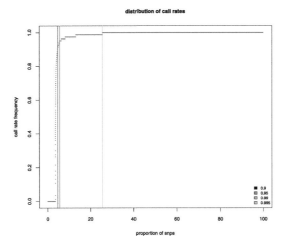

Figure 3: distribution of call rates per SNP.

band (see breakdown of call rates in 2). Note that this will not hold well if there is ascertainment bias problems with the SNPs (i.e. SNPs selected for the chip derived from one population and the samples come from a very different one). In this dataset 3125 SNPs failed genotyping in over 5% of the samples (these were removed from the dataset). Note that the number of SNPs failed depends on the GC cutoff threshold – all SNPs below 0.5 are deemed to have failed (see further details in GC scores section).

2.2   GC scores

GC scores were filtered for a threshold value of 0.5. All calls under this value were discarded (note that this is specific for each snp on an individual sample). The dataset contained 650 SNPs where all GC scores were 0. A further 2557 SNPs had a GC score over 0.5 in less than 90% of the samples. 30076 SNPs had a GC score of at least 0.9 for at least 90% of the genotypes. The mean GC

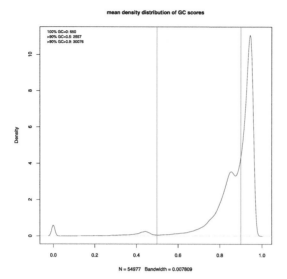

Figure 4: Histogram of GC scores.

scores for this data is 0.859 and the median is 0.865. The distribution of GC scores is shown in Figures 4 and 5.

2.3   Minor allele frequency

The minor allele frequency (MAF) was calculated for each SNP. 178 SNPs are homozygous for the locus. A further 129 had a MAF below 0.01 and were discarded. The distribution of MAFs is shown in figure 6. The average heterozygosity for the SNPs is 0.39 and the standard deviation is 0.137. A total 6 SNPs were detected as outliers (3SD from the mean and removed). Heterozygosity (He) and gene diversity (Ho) distributions are shown in figure 7.

2.4   Hardy-Weinberg equilibrium

Hardy-Weinberg (HW) equilibrium was calculated for each individual SNP using an exact chi-square test with continuity correction. HW equilibirum could not be determined for 2210 SNPS because these were either homozygous or had no calls assigned. 127 SNPs had a p-value of 0. A p-value cutoff of 1e-15 shows 131 SNPs out of HW equilibrium (note that this also includes SNPs that would not be expected to be in HW equilibrium such as those on sex chromosomes, mitochondria, etc). Figure 8 shows the distribution of p-values for HW equilibrium.

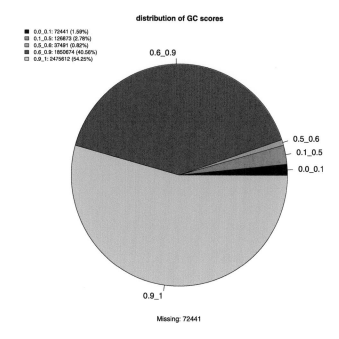

Figure 5: Pie plot of GC scores.

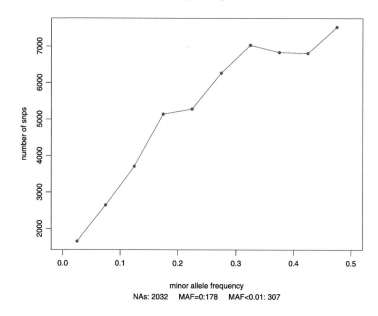

Figure 6: Minor allele frequency distribution for SNPs.

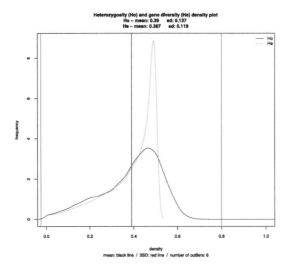

Figure 7: Heterozygosity distribution for SNPs. Note: standard deviations are biased.

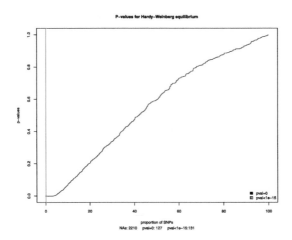

Figure 8: P-value distribution and thresholds for Hardy-Weinberg equilibrium.

7

References

1. ADLER, J. *R in a Nutshell*. O'Reilly, 2009.
2. AFFYMETRIX. Statistical algorithms description document. Tech. rep., Affymetrix, 2002.
3. AFFYMETRIX. Guide to probe logarithmic intensity error (plier) estimation. Tech. rep., Affymetrix, 2005.
4. ALBERT, J. *Bayesian Computation with R*. Springer, New York, 2007.
5. ALBRECHTSEN, A., NIELSEN, F. C., AND NIELSEN, R. Ascertainment biases in snp chips affect measures of population divergence. *Mol Biol Evol 27*, 11 (2010), 2534–47.
6. ALEXA, A., RAHNENFUHRER, J., AND LENGAUER, T. Improved scoring of functional groups from gene expression data by decorrelating go graph structure. *Bioinformatics 22* (2006), 1600–1607.
7. ANDERS, S., MCCARTHY, D. J., CHEN, Y., OKONIEWSKI, M., SMYTH, G. K., HUBER, W., AND ROBINSON, M. D. Count-based differential expression analysis of rna sequencing data using r and bioconductor. *Nat Protoc 8*, 9 (2013), 1765–86.
8. AUER, P. L., AND DOERGE, R. W. Statistical design and analysis of rna sequencing data. *Genetics 185*, 2 (2010), 405–16.
9. BACLAWSKI, K. P. *Introduction to Probability with R*. Chapman & Hall/CRC, Boca Raton, FL, 2008.
10. BALL, R. D. Designing a gwas: power, sample size, and data structure. *Methods in Molecular Biology 1019* (2013), 37–98.
11. BESSANT, C., OAKLEY, D., AND SHADFORTH, I. *Building Bioinformatics Solutions*. Oxford University Press, 2014.
12. BOLGER, A. M., LOHSE, M., AND USADEL, B. Trimmomatic: a flexible trimmer for illumina sequence data. *Bioinformatics 30*, 15 (2014), 2114–20.
13. BUSH, W. S., AND MOORE, J. H. Chapter 11: Genome-wide association studies. *PLoS Computational Biology 8*, 12 (2012), e1002822.
14. CARLIN, B. P., AND LOUIS, T. A. *Bayesian Methods for Data Analysis*. Chapman & Hall/CRC, Boca Raton, FL, 2008.
15. CASELLA, C. P. R. G. *Introducing Monte Carlo Methods with R*. Springer, 2010.
16. CHAMBERS, J. M. *Software for Data Analysis: Programming with R*. Springer, New York, 2008.
17. COHEN, Y., AND COHEN, J. Y. *Statistics and Data with R*. Wiley, 2008.
18. CONSORTIUM, T. G. O. Gene ontology: tool for the unification of biology. *Nature Genetics 25* (2000), 25–29.
19. COOK, D., AND SWAYNE, D. F. *Interactive and Dynamic Graphics for Data Analysis*. Springer, New York, 2007.

© Springer International Publishing Switzerland 2015
C. Gondro, *Primer to Analysis of Genomic Data Using R*, Use R!,
DOI 10.1007/978-3-319-14475-7

20. CRAWLEY, M. J. *Statistics: An Introduction using R*. Wiley, Chichester, UK, 2005.

21. CRAWLEY, M. J. *The R Book*. Wiley, Chichester, UK, 2012.

22. DALGAARD, P. *Introductory Statistics with R*, 2nd ed. Springer, 2008.

23. DILLIES, M. A., RAU, A., AUBERT, J., HENNEQUET-ANTIER, C., JEANMOUGIN, M., SERVANT, N., KEIME, C., MAROT, G., CASTEL, D., ESTELLE, J., GUERNEC, G., JAGLA, B., JOUNEAU, L., LALOE, D., LE GALL, C., SCHAEFFER, B., LE CROM, S., GUEDJ, M., JAFFREZIC, F., AND FRENCH STATOMIQUE, C.   A comprehensive evaluation of normalization methods for illumina high-throughput rna sequencing data analysis. *Brief Bioinform 14*, 6 (2013), 671–83.

24. DOBSON, A. J., BARNETT, A., AND GROVE, K. *An Introduction to Generalized Linear Models*. Chapman & Hall/CRC, Boca Raton, FL, 2008.

25. DUDOIT, S., AND VAN DER LAAN, M. J. *Multiple Testing Procedures with Applications to Genomics*. Springer, New York, 2008.

26. EDDELBUETTEL, D. *Seamless R and C++ Integration with Rcpp*. Springer, 2013.

27. EUBANK, R. L., AND KUPRESANIN, A. *Statistical Computing in C++ and R*. CRC Press, 2011.

28. EUGSTER, M., KNAUS, J., PORZELIUS, C., SCHMIDBERGER, M., AND VICEDO, E. Hands-on tutorial for parallel computing with r. *Computational Statistics 26* (2011), 219–239.

29. EVERITT, B., AND HOTHORN, T. *A Handbook of Statistical Analyses Using R*. Chapman & Hall/CRC, Boca Raton, FL, 2006.

30. FALCONER, D. S., AND MACKAY, T. F.   *Introduction to Quantitative Genetics*, 4th ed. Benjamin Cummings, 1996.

31. FARAWAY, J. J. *Linear Models with R*. Chapman & Hall/CRC, Boca Raton, FL, 2004.

32. FARAWAY, J. J.   *Extending Linear Models with R: Generalized Linear, Mixed Effects and Nonparametric Regression Models*. Chapman & Hall/CRC, Boca Raton, FL, 2006.

33. FAY, J. C., AND WU, C.-I. Hitchhiking under positive Darwinian selection. *Genetics 155*, 3 (2000), 1405–1413.

34. FERNANDO, R. L., AND GARRICK, D. Bayesian methods applied to gwas. In *Genome-Wide Association Studies and Genomic Prediction*. Springer, 2013, pp. 237–274.

35. FOULKES, A. S. *Applied Statistical Genetics with R: For Population-based Association Studies*. Springer, New York, 2009.

36. FOX, J. The r commander: A basic-statistics graphical user interface to r. *Journal of Statistical Software 14*, 9 (8 2005), 1–42.

37. GAUTIER, L., MOOLLER, M., FRIIS-HANSEN, L., AND KNUDSEN, S. Alternative mapping of probes to genes for affymetrix chips. *BMC Bioinformatics 5* (2004), 111.

38. GAUTIER, M., AND VITALIS, R. rehh: an r package to detect footprints of selection in genome-wide snp data from haplotype structure. *Bioinformatics 28*, 8 (2012), 1176–1177.

39. GENTLEMAN, R. *R Programming for Bioinformatics*, vol. 12 of *Computer Science & Data Analysis*. Chapman & Hall/CRC, Boca Raton, FL, 2008.

40. GENTLEMAN, R., CAREY, V., HUBER, W., IRIZARRY, R., AND DUDOIT, S., Eds. *Bioinformatics and Computational Biology Solutions Using R and Bioconductor*. Statistics for Biology and Health. Springer, 2005.

41. GODDARD, M. Genomic selection: prediction of accuracy and maximisation of long term response. *Genetica 136*, 2 (2009), 245–57.

42. GONDRO, C., AND KINGHORN, B. P. Optimization of cDNA microarray experimental designs using an evolutionary algorithm. *IEEE/ACM Trans Comput Biol Bioinform 5*, 4 (2008), 630–638.

43. GONDRO, C., LEE, S. H., LEE, H. K., AND PORTO-NETO, L. R. Quality control for genome-wide association studies. *Methods Mol Biol 1019* (2013), 129–47.

44. GONDRO, C., PORTO-NETO, L. R., AND LEE, S. H. R for genome-wide association studies. *Methods Mol Biol 1019* (2013), 1–17.

45. GONDRO, C., PORTO-NETO, L. R., AND LEE, S. H. snpqc - an r pipeline for quality control of illumina snp genotyping array data. *Anim Genet 45*, 5 (2014), 758–61.

46. GONDRO, C., VAN DER WERF, J., AND HAYES, B. J. *Genome-wide Association Studies and Genomic Prediction*. Springer, 2013.

47. HAHNE, F., HUBER, W., GENTLEMAN, R., AND FALCON, S. *Bioconductor Case Studies*. Springer, New York, 2008.

48. HARDIMAN, G. Microarray platforms - comparisons and contrasts. *Pharmacogenomics 5*, 5 (2004), 487–502.

49. HARRISON, A., JOHNSTON, C., AND ORENGO, C. Establishing a major cause of discrepancy in the calibration of affymetrix genechips. *BMC Bioinformatics 8* (2007), 195.

50. HART, S. N., THERNEAU, T. M., ZHANG, Y., POLAND, G. A., AND KOCHER, J. P. Calculating sample size estimates for rna sequencing data. *J Comput Biol 20*, 12 (2013), 970–8.

51. HARTL, D. L., AND CLARK, A. *Principles of Population Genetics*. Sinauer, 2007.

52. HAYES, B. Overview of statistical methods for genome-wide association studies (gwas). *Methods Mol Biol 1019* (2013), 149–169.

53. HAYES, B. J., CHAMBERLAIN, A. J., MACEACHERN, S., SAVIN, K., MCPARTLAN, H., MACLEOD, I., SETHURAMAN, L., AND GODDARD, M. E. A genome map of divergent artificial selection between bos taurus dairy cattle and bos taurus beef cattle. *Animal Genetics 40*, 2 (2009), 176–184.

54. HOLSINGER, K. E., AND WEIR, B. S. Genetics in geographically structured populations: defining, estimating and interpreting fst. *Nature Reviews Genetics 10*, 9 (2009), 639–650.

55. HUANG, X., AND HAN, B. Natural variations and genome-wide association studies in crop plants. *Annu Rev Plant Biol 65* (2014), 531–51.

56. HUBER, W., VON HEYDEBRECK, A., SULTMANN, H., POUSTKA, A., AND VINGRON, M. Variance stabilization applied to microarray data calibration and to quantification of differential expression. *Bioinformatics 18* (2002), S96–S104.

57. IRIZARRY, R., BOLSTAD, B., COLLIN, F., COPE, L., HOBBS, B., AND SPEED, T. Summaries of affymetrix genechip probe level data. *Nucleic Acids Research 31* (2003), e15.

58. IRIZARRY, R., HOBBS, B., COLLIN, F., BEAZER-BARCLAY, Y., ANTONELLIS, K., SCHERF, U., AND SPEED, T. Exploration, normalization, and summaries of high density oligonucleotide array probe level data. *Biostatistics 4* (2003), 249–264.

59. IRIZARRY, R., WU, Z., AND JAFFE, H. Comparison of affymetrix genechip expression measures. *Bioinformatics 22* (2006), 789–794.

60. JONES, O., MAILLARDET, R., AND ROBINSON, A. *Scientific Programming and Simulation using R*. CRC Press, 2009.

61. KEEN, K. J. *Graphics for Statistics and Data Analysis with R*. Chapman and Hall/CRC, 2010.

62. KLEIN, R. J. Power analysis for genome-wide association studies. *BMC Genetics 8* (2007), 58.

63. KRUGLYAK, L. The road to genome-wide association studies. *Nature Reviews Genetics 9*, 4 (2008), 314–318.

64. LACHANCE, J., AND TISHKOFF, S. A. Snp ascertainment bias in population genetic analyses: Why it is important, and how to correct it. *Bioessays 35*, 9 (2013), 780–6.

65. LANDSMAN, D., GENTLEMAN, R., KELSO, J., AND OUELLETTE, B. F. F. Database: A new forum for biological databases and curation. *DATABASE 2009* (2009), bap002.

66. LANGMEAD, B., AND SALZBERG, S. L. Fast gapped-read alignment with bowtie 2. *Nat Methods 9*, 4 (2012), 357–9.

67. LI, C., AND WONG, W. Model-based analysis of oligonucleotide arrays: Expression index computation and outlier detection. *PNAS 98* (2001), 31–36.

68. LI, H., HANDSAKER, B., WYSOKER, A., FENNELL, T., RUAN, J., HOMER, N., MARTH, G., ABECASIS, G., DURBIN, R., AND GENOME PROJECT DATA PROCESSING, S. The sequence alignment/map format and samtools. *Bioinformatics 25*, 16 (2009), 2078–9.

69. LIU, Y., ZHOU, J., AND WHITE, K. P. Rna-seq differential expression studies: more sequence or more replication? *Bioinformatics 30*, 3 (2014), 301–4.

70. MAINDONALD, J., AND BRAUN, J. *Data Analysis and Graphics Using R*, 3rd ed. Cambridge University Press, Cambridge, 2010.

71. MARSHALL, K., MADDOX, J., LEE, S., ZHANG, Y., KAHN, L., GRASER, H., GONDRO, C., WALKDEN-BROWN, S., AND VAN DER WERF, J. H. Genetic mapping of quantitative trait loci for resistance to haemonchus contortus in sheep. *Animal Genetics 40*, 3 (2009), 262–272.

72. MATSUMURA, H., KRUGER, D. H., KAHL, G., AND TERAUCHI, R. Supersage: a modern platform for genome-wide quantitative transcript profiling. *Curr Pharm Biotechnol 9*, 5 (2008), 368–74.

73. MATSUMURA, H., URASAKI, N., YOSHIDA, K., KRUGER, D. H., KAHL, G., AND TERAUCHI, R. Supersage: powerful serial analysis of gene expression. *Methods Mol Biol 883* (2012), 1–17.

74. McCALLUM, E., AND WESTON, S. *Parallel R*. O'Reilly Media, Inc, 2011.

75. MEUWISSEN, T., HAYES, B., AND GODDARD, M. Prediction of total genetic value using genome-wide dense marker maps. *Genetics 157*, 4 (2001), 1819–1829.

76. MORGAN, M., ANDERS, S., LAWRENCE, M., ABOYOUN, P., PAGES, H., AND GENTLEMAN, R. Shortread: a bioconductor package for input, quality assessment and exploration of high-throughput sequence data. *Bioinformatics 25*, 19 (2009), 2607–8.

77. MRODE, R. A. *Linear models for the prediction of animal breeding values*. Cabi, 2014.

78. MURRELL, P. *R Graphics*. Chapman & Hall/CRC, Boca Raton, FL, 2005.

79. NEALE, B. M., AND PURCELL, S. The positives, protocols and perils of genome-wide association. *American Journal of Medical Genetics Part B 147B*, 7 (2008), 1288–1294.

80. NEI, M. Analysis of gene diversity in subdivided populations. *Proceedings of the National Academy of Sciences 70*, 12 (1973), 3321–3323.

81. NICHOLSON, G., SMITH, A. V., JÓNSSON, F., GÚSTAFSSON, Ó., STEFÁNSSON, K., AND DONNELLY, P. Assessing population differentiation and isolation from single-nucleotide polymorphism data. *Journal of the Royal Statistical Society: Series B (Statistical Methodology) 64*, 4 (2002), 695–715.

82. OSHLACK, A., ROBINSON, M. D., AND YOUNG, M. D. From rna-seq reads to differential expression results. *Genome Biol 11*, 12 (2010), 220.

83. PARADIS, E. pegas: an r package for population genetics with an integrated–modular approach. *Bioinformatics 26*, 3 (2010), 419–420.

84. PARADIS, E. *Analysis of Phylogenetics and Evolution with R*, 2nd ed. Use R. Springer, New York, 2012.

85. PATEL, J. N., McLEOD, H. L., AND INNOCENTI, F. Implications of genome-wide association studies in cancer therapeutics. *British Journal of Clinical Pharmacology 76*, 3 (2013), 370–380.

86. PINHEIRO, J. C., AND BATES, D. M. *Mixed-Effects Models in S and S-Plus*. Springer, 2000.

87. PIPER, E., JONSSON, N., GONDRO, C., LEW-TABOR, A., MOOLHUIJZEN, P., ME, M. V., AND JACKSON, L. Immunological profiles of bos taurus and bos indicus cattle infested with the cattle tick, rhipicephalus (boophilus) microplus. *Clinical and Vaccine Immunology (epub ahead of print)* (2009).

88. PITT-FRANCIS, J., AND WHITELEY, J. *Guide to scientific computing in C++*. Springer, 2012.

89. PORTO-NETO, L. R., LEE, S. H., LEE, H. K., AND GONDRO, C. Detection of signatures of selection using fst. In *Genome-Wide Association Studies and Genomic Prediction*. Springer, 2013, pp. 423–436.

90. R DEVELOPMENT CORE TEAM. *R: A Language and Environment for Statistical Computing*. R Foundation for Statistical Computing, Vienna, Austria, 2014.

91. RAMALHO, J. A. *Learn SQL*. Wordware Publishing, Plano, USA, 2000.

92. RAPAPORT, F., KHANIN, R., LIANG, Y. P., PIRUN, M., KREK, A., ZUMBO, P., MASON, C. E., SOCCI, N. D., AND BETEL, D. Comprehensive evaluation of differential gene expression analysis methods for rna-seq data. *Genome Biology 14*, 9 (2013).

93. RINCON, G., WEBER, K. L., EENENNAAM, A. L., GOLDEN, B. L., AND MEDRANO, J. F. Hot topic: performance of bovine high-density genotyping platforms in Holsteins and Jerseys. *J Dairy Sci 94*, 12 (2011), 6116–21.

94. RIZZO, M. L. *Statistical Computing with R*. Chapman & Hall/CRC, Boca Raton, FL, 2008.

95. SABETI, P., SCHAFFNER, S., FRY, B., LOHMUELLER, J., VARILLY, P., SHAMOVSKY, O., PALMA, A., MIKKELSEN, T., ALTSHULER, D., AND LANDER, E. Positive natural selection in the human lineage. *science 312*, 5780 (2006), 1614–1620.

96. SARKAR, D. *Lattice Multivariate Data Visualization with R.* Springer, New York, 2007.

97. SCHENA, M., SHALON, D., DAVIS, R., AND BROWN, P. Quantitative monitoring of gene expression patterns with complementary DNA microarray. *Science 270* (1995), 467–470.

98. SCHMIDBERGER, M., MORGAN, M., EDDELBUETTEL, D., YU, H., AND MANSMANN, L. T. U. State of the art in parallel computing with r. *Journal of Statistical Software 31(1)* (2009).

99. SCHULZE, A., AND DOWNWARD, J. Navigating gene expression using microarrays - a technology review. *Nature Cell Biology 3*, 8 (2001), E190–E195.

100. SHEATHER, S. *A Modern Approach to Regression with R.* Springer, New York, 2009.

101. SHENDURE, J. The beginning of the end for microarrays? *Nature Methods 5* (2008), 585–587.

102. SIEGMUND, D., AND YAKIR, B. *The Statistics of Gene Mapping.* Springer, New York, 2007.

103. SLONIM, D. K., AND YANAI, I. Getting started in gene expression microarray analysis. *PLoS Computational Biology 5*, 10 (2009).

104. SMYTH, G. K. Linear models and empirical Bayes methods for assessing differential expression in microarray experiments. *Statistical Applications in Genetics and Molecular Biology 3*, 1 (2004), 3.

105. SPECTOR, P. *Data Manipulation with R.* Springer, New York, 2008.

106. SPENCER, C. C. A., SU, Z., DONNELLY, P., AND MARCHINI, J. Designing genome-wide association studies: Sample size, power, imputation, and the choice of genotyping chip. *PLoS Genetics 5*, 5 (2009), e1000477.

107. SUESS, E. A., AND TRUMBO, B. E. *Introduction to Probability Simulation and Gibbs Sampling with R.* Springer, 2010.

108. TAJIMA, F. Statistical method for testing the neutral mutation hypothesis by DNA polymorphism. *Genetics 123*, 3 (1989), 585–595.

109. TEO, Y. Y. Common statistical issues in genome-wide association studies: a review on power, data quality control, genotype calling and population structure. *Current Opinion in Lipidology 19*, 2 (2008), 133–143.

110. TIERNEY, L. *Compstat 2008.* Springer, New York, 2008, ch. Implicit and Explicit Parallel Computing in R, pp. 43–51.

111. TIERNEY, L., ROSSINI, A., AND LI, N. Snow : A parallel computing framework for the r system. *International Journal of Parallel Programming 37*, 1 (2009), 78–90.

112. UGARTE, M. D., MILITINO, A. F., AND ARNHOLT, A. *Probability and Statistics with R.* Chapman & Hall/CRC, Boca Raton, FL, 2008.

113. VANRADEN, P. Efficient methods to compute genomic predictions. *Journal of dairy science 91*, 11 (2008), 4414–4423.

114. VERZANI, J. *Using R for Introductory Statistics.* Chapman & Hall/CRC, Boca Raton, FL, 2005.

115. WANG, Z., GERSTEIN, M., AND SNYDER, M. Rna-seq: a revolutionary tool for transcriptomics. *Nature Reviews Genetics 10*, 1 (2009), 57–63.

116. WEHRENS, R. *Chemometrics with R: multivariate data analysis in the natural sciences and life sciences.* Springer, 2011.

117. WEIR, B. S., AND COCKERHAM, C. C. Estimating f-statistics for the analysis of population structure. *evolution* (1984), 1358–1370.

118. WOO, Y., AFFOURTIT, J., DAIGLE, S., VIALE, A., JOHNSON, K., NAGGERT, J., AND CHURCHILL, G. A comparison of cDNA, oligonucleotide, and affymetrix genechip gene expression microarray platforms. *J Biomol Tech 15*, 4 (2004), 276–84.

119. WRIGHT, S. The genetical structure of populations. *Annals of Eugenics 15* (1951), 323–354.

120. WU, R., MA, C., AND CASELLA, G. *Statistical Genetics of Quantitative Traits: Linkage, Maps and QTL.* Springer, New York, 2007.

121. WU, Z., RA, I., GENTLEMAN, R., MURILLO, F. M., AND SPENCER, F. A model based background adjustment for oligonucleotide expression arrays. *Journal of the American Statistical Association 99* (2003), 909–917.

122. XU, X., ZHANG, Y., WILLIAMS, J., ANTONIOU, E., McCOMBIE, W. R., WU, S., ZHU, W., DAVIDSON, N. O., DENOYA, P., AND LI, E. Parallel comparison of illumina rna-seq and affymetrix microarray platforms on transcriptomic profiles generated from 5-aza-deoxy-cytidine treated ht-29 colon cancer cells and simulated datasets. *BMC Bioinformatics 14 Suppl 9* (2013), S1.

123. ZHANG, A. *Advanced analysis of gene expression microarray data.* World Scientific, London, UK, 2006.

124. ZHAO, S., FUNG-LEUNG, W. P., BITTNER, A., NGO, K., AND LIU, X. Comparison of rna-seq and microarray in transcriptome profiling of activated t cells. *PLoS One 9*, 1 (2014), e78644.

125. ZIEGLER, A., KONIG, I. R., AND THOMPSON, J. R. Biostatistical aspects of genome-wide association studies. *Biometrical Journal of Statistical Software 50*, 1 (2008), 8–28.

126. ZUUR, A. F., IENO, E. N., AND MEESTERS, E. *A Beginner's Guide to R.* Use R. Springer, 2009.

CPI Antony Rowe
Eastbourne, UK
August 06, 2019